A Tour of the Flowering Plants

Based on the Classification System of the Angiosperm Phylogeny Group

A Tour of the Flowering Plants

Based on the Classification System of the Angiosperm Phylogeny Group

Text and photographs by
PRISCILLA SPEARS

🌱 MISSOURI BOTANICAL GARDEN PRESS

A Tour of the Flowering Plants
Based on the Classification System of the Angiosperm Phylogeny Group

Front cover: white evening primroses, *Oenothera albicaulis*

Back cover: Top left: Colorado columbine, *Aquilegia caerulea*; right: Asian lily hybrid, *Lilium*;
bottom: water lily hybrid, *Nymphaea*

Frontispiece: pasqueflowers, *Pulsatilla patens*

Facing page 1: Foreground: monkshood, *Aconitum columbianum;*
background: corn husk lily, *Veratrum tenuipetalum*

ISBN 1-930723-48-2

Library of Congress Control Number: 2006921483

Scientific Editor and Head,
Missouri Botanical Garden Press: **Victoria C. Hollowell**

Managing Editor: **Beth Parada**

Associate Editor: **Diana Gunter**

Editorial Assistant: **Barbara Mack**

Press Assistant: **Adele Niblack**

Cover design: Gobberdiel Graphic Design, St. Louis, Missouri

To James A. Brierley, educator

Contents

Foreword

In 1993, a landmark paper by Mark Chase and 41 coauthors appeared in the *Annals of the Missouri Botanical Garden,* in which an outline of the phylogeny of flowering plants was sketched. More work, mostly collaborative, followed fast and furious, and by 1998 the outline of relationships appeared stable enough to propose a revised classification of flowering plants. Its guiding principle is monophyly, "that taxonomic groups should hold all the descendants of a single ancestor, and only its descendants," as Priscilla Spears explains. *A Tour of the Flowering Plants* is based on the only slightly changed successor to the first classification, which appeared in 2003. Indeed, if outlines of the phylogeny remain unchanged, there will be no need for the classification to change further. Textbooks at all levels, floras, and even the arrangement of herbaria and botanical gardens around the world are being made compatible with this new classification. In *A Tour of the Flowering Plants*, Priscilla Spears introduces the plants of much of the United States and Canada in this new context. The sheer beauty and botanical detail illustrated in the hundreds of photographs that grace this book recommend it to the professional botanist and general plant-lover alike, and the text is a goldmine of all sorts of botanical information.

Peter Stevens
University of Missouri, St. Louis,
and Missouri Botanical Garden

Acknowledgments

Many people assisted me on my journey of preparing this book. Carolyn Jones of InPrint for Children, in Glenside, Pennsylvania, helped me work out the page design and was a good listener for the whole process. She graciously contributed the drawing for the parts of the flower that enhances the introduction. The photos of the *Magnolia grandiflora* flower, the North American tulip tree, the lizard tail inflorescence, the mountain laurel flowers, and the overviews of *Cornus florida* and *C. kousa* are also generous gifts from Carolyn and are used with her permission. Lisa Gould of the Rhode Island Natural History Survey, Inc. kindly supplied her photo of *Sassafras* flowers and Irene H. Stuckey's photo of *Lindera* berries. Lisa also introduced me to the flora of her state on a personal field trip. Jay Sullivan extended kindness to a stranger and gave me permission to use his photo of *Anemopsis*. All other photos are my own work.

Anna Kalkwarf helped me find plants, which she does very well, and assisted me on photography expeditions. She was a patient hiking partner as I photographed. She also shared her garden and houseplants with me, providing several subjects for photos. Several other friends opened their gardens to me, principally Karla Briggs, who supplied heartening encouragement as well as photography subjects from her garden and greenhouse. I also am grateful for the opportunity to photograph in the gardens of Julaine Kennedy, Lynn Allbright, and Marilyn Flanigan. Holly Jarvis of Waco, Texas, went out of her way to help me find photography subjects when I visited her area in July 2004. I thank Colorado botanist Bill Jennings for helping me identify the native orchids in my photos.

This book would have been impossible without the many plant resources of the Denver Botanic Gardens. I found many treasures to photograph as I poked in all the corners. The personnel were a wonderful help. Joe Tomachek, curator of the water gardens, took time to help me with *Acorus* specimens and tutored me on cattails. John Bayard, of the International Carnivorous Plant Society, graciously gave me a tour of the gorgeous collection of insectivorous plants. I am grateful for the opportunity to photograph plants in the institutional greenhouses. In the Kathryn Kalmbach Herbarium, Janet Wingate and Loraine Yates assisted me with identification of several grasses and native eudicots. I thank the many courteous personnel of the Denver Botanic Gardens who answered my questions. Special thanks go to Alison Graber in the Helen Fowler Library, who thought she was printing out the reference page, but actually printed out the full 59-page Angiosperm Phylogeny Group 2003 paper, which inspired me to tackle this project.

The Hudson Gardens of Littleton, Colorado, provided me with many photographic subjects. I am especially appreciative of their spectacular lotus pond. The arboretum at Regis University in Denver was another source of my photographic subjects. Tatonka Farms of Conifer, Colorado, and Stems: A Flower Shop of Evergreen, Colorado, allowed me to photograph their merchandise.

Other institutions where I photographed for this book include: San Francisco Botanical Garden at Strybing Arboretum and the Conservatory of Flowers in Golden Gate Park; Balboa Park in San Diego; Bellevue Botanical Gardens in Bellevue, Washington; the Arizona-Sonora Desert Museum, Tucson; and the Tucson Botanical Gardens.

I have been inspired by the public lectures of Ned Friedmann of the University of Colorado. Ned set a great example for bridging between cutting-edge research and the general public with his talks at the Denver Museum of Nature and Science. He kindly looked over the first rough draft of this book and made several helpful suggestions.

I am most grateful to Peter Stevens of the Missouri Botanical Garden and of the University of Missouri, St. Louis, the keeper of the Angiosperm Phylogeny Website, for his review of the manuscript both at first draft and later stages. This book would not exist without Peter's help. He made many helpful comments, from which I learned a great deal and which enabled me to significantly improve the book. His support of this project has been crucial, especially since he introduced the manuscript to Victoria Hollowell, scientific editor and head of the Missouri Botanical Garden Press.

Victoria helped me turn an incomplete first draft into what I had hoped, a useful reference book. She guided me through the sometimes bewildering world of scientific nomenclature and terminology, and she helped me make the work more accurate and useable by the advanced part of the audience, as well as by beginners. Barbara Alongi read the first draft and made helpful suggestions for further illustrations.

Beth Parada of the Missouri Botanical Garden Press performed the crucial task of turning the manuscript into a published book, no mean feat for a book like this. I appreciate her keen eye for typos and inconsistencies, as well as her handling of technical details.

My family has been a great support. Special thanks go to my father, Charles H. Spears, for building me a very useful rolling bookcase, without which I would likely be under a collapsed stack of reference books. My husband, Daniel Bean, provided support in many ways, from technical advice on computer graphics to reaching high branches and bringing them down to my camera.

In spite of all this help, the mistakes in this work are mine, and I would appreciate knowing about them, so that I can correct them in future editions. Please send your comments to me via Missouri Botanical Garden Press.

Introduction

Welcome to *A Tour of the Flowering Plants*

Why take this tour? Do you enjoy attractive color and beautiful form? How about interesting life stories? Do you want to know more about the plants around you? Do you want to know more about what makes our planet and its life work? This tour has something for you!

Life on land depends on the flowering plants, the angiosperms, as primary producers and the foundation of most ecosystems. Angiosperms play important roles in aquatic ecosystems as well. Beyond their vital ecological roles, flowering plants are beautiful and visually interesting, with a wide variety of forms and life stories. One shouldn't miss out on these fascinating organisms. Many people, however, are able to name only a few plants around them. Many cannot see the amazing diversity of flowers because they have never learned to look. I hope this tour will open many eyes to the beauty and wonderful stories of the flowering plants and, at the same time, increase understanding of the diversity of this vital part of the living world.

Where we will tour: This book, like a real tour, will give you a quick look at a wide spectrum of its subject area, the flowering plants in temperate areas of the United States and Canada. It provides an overview of their families and evolutionary relationships, and introduces an up-to-date system of classification. At the end, I hope you will be inspired to observe more plants around you and find their place in the flowering plants. For more in-depth information about flowering plant families, see the Selected References section at the end of the book.

What families we will see: The 109 families of flowering plants that are illustrated here represent well-known or common families in the United States and Canada, ones that we use for food or medicine or grow in our gardens. Some families are shown because they are important branches on the "family tree" of flowering plants. Because the flower itself is the major distinguishing feature of the families, I mainly illustrate the flowers, with photos of fruits and other plant parts as supplements. For a more extensive outline of flowering plant families, refer to Appendix A. This listing will also help you to see the structure of the Angiosperm Phylogeny Group classification system.

Why this tour uses a recent classification: The science that studies the diversity of life, systematics, has advanced greatly in the past two decades. We have traditionally placed organisms that resemble one another in groups, but now we are also concerned with the relationships among these groups. Current scientific thought maintains that taxonomic groups should hold all the descendants of a single ancestor, and only its descendants. Recent systematics studies use all the information we had previously on the structure of plants, and add additional information about the DNA and the development of plants. The result is similar to traditional classification, but with important differences. The remodeled system reflects the evolutionary history of the organisms.

The revised classification for flowering plants, also known as the angiosperms, has been and continues to be developed by a team of scientists called the Angiosperm Phylogeny Group (APG). Members of this large, worldwide team have pooled their data and efforts to produce a new picture of flowering plant relationships. The result is a phylogenetic classification. "Phylogeny" refers to the evolutionary history of a group of organisms. Phylogenetic classifications show the branching of the tree of life and reflect the evolutionary history of life.

A map to help you find your way around the flowering plants. The tree diagram below is our map for the tour. It shows the basic relationships among the orders of flowering plants. You will see this map again at the introduction to certain groups, with highlighting to show the upcoming sections.

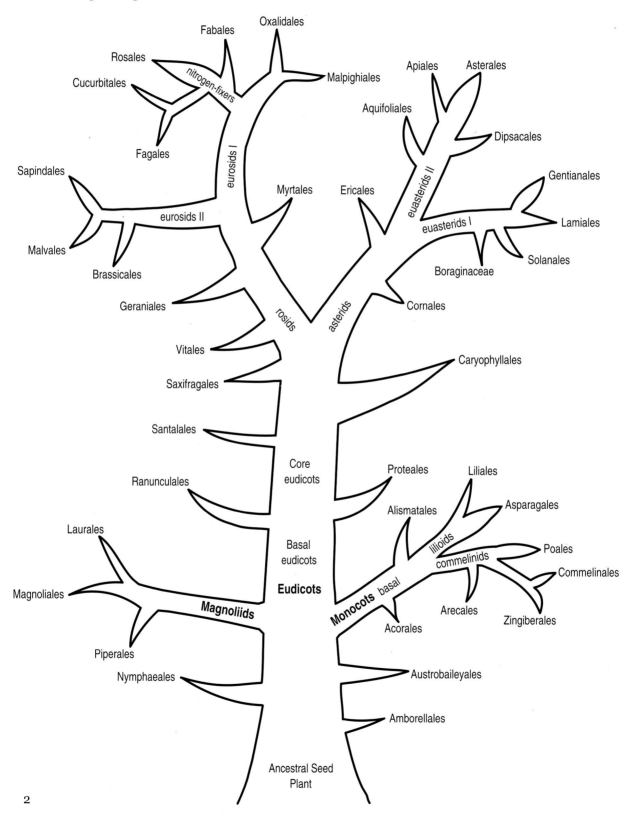

The real maps of the flowering plants used by botanists are much more complex and quantitative. Recent botanical publications and websites show a wealth of such phylogenetic diagrams. Cladograms are one type of branching diagram. See the Selected References section if you are interested in further explorations of branching diagrams.

How our tour operates: The colored labels on the right side of the page will help you navigate this book. The top label and the color divide the flowering plants into three parts—basal angiosperms, monocots, and eudicots. The second label gives a subgrouping, such as eurosids II or core eudicots. These lineages are not given a traditional hierarchical category such as subclass or superorder. You can see the middle label subgroups on our tree diagram. From these subgroups down, our classification looks much like traditional systems. There are orders, families, genera, and species. The bottom label on the page lists the order shown. The label across the top of the page lists the family, as well as subfamily or tribe as needed. You will be able to recognize plant order names, since they always end in –ales. Family names end in –aceae.

The flowering plant orders are presented according to their place on the tree diagram. Within each order, the families are placed in alphabetical order. It would be nice to have the families in phylogenetic order, but these relationships are not yet known well enough for every family in our tour. For more information on the phylogeny of families within the orders, see Principal Resources in the Selected References section at the end of the book.

For each family, I've listed the classification on the left page across from its pictures. This includes major genera in the family, along with common names. I've given the approximate number of genera and species for that family *as it occurs worldwide*. An asterisk marks the names of plants species that are illustrated in our tour. If it was not possible to identify a plant to species, the asterisk marks the genus of the plant illustrated.

The green labels mark the basal angiosperms. This group makes up about only about 3% of the flowering plants, but it includes some intriguing lines of descent. The term "basal" means "at the base" and commonly refers to a group or groups that diverge early from a lineage. Basal lineages, like any other lineage, are not static—they continue on their own evolutionary path. On our tree diagram of flowering plant orders, my basal angiosperms are the magnoliid branch and the three lowest branches. Plants of this group have some features in common with monocots and eudicots, but they belong to neither group. For now, I place them in their own "miscellaneous drawer."

The monocots have purple labels. Monocots are all one related group, members of a single lineage. Their subgroups are the basal monocots, the lilioid or petalloid monocots, and the commelinid monocots. Monocots make up about a quarter of the flowering plants.

The eudicots have blue labels. The traditional class for dicots included the basal angiosperms. With the basal groups subtracted out, the remainder is called the **eudicots**, meaning "true dicots" and pronounced "U-dicots." They are the largest group of flowering plants, making up over 70% of the quarter-million-plus species. They are divided further into basal eudicots and core eudicots. Within core eudicots, there are five main groups—Caryophyllales, Santalales, Saxifragales, the rosids, and the asterids. The largest of these groups are the rosids and asterids, each of which have two main branches. The subgroups are called eurosids I and II, and euasterids I and II.

Changes of note in the classification of flowering plants:

The recognition of basal angiosperms makes the class Magnoliopsida, the dicotyledons or dicots, obsolete. There are no longer two classes of angiosperms. The magnoliids and the water lilies are not closely related to the eudicots, and therefore cannot share their class. This situation, like so much of life, isn't simple. Our map, the tree diagram, shows this. This diagram is simplified; it mainly shows orders from the United States and Canada that are featured on this tour. The full tree for flowering plants would have many more branches.

Recent changes in the monocots include placing sweet flag, *Acorus*, as the most basal monocot. The main groups of monocots are the basal monocots, the lilioid or petaloid monocots (botanists use both terms so I include them both), and the commelinid monocots. On the family level, the break-up of the lily family is a big change. Field guides and gardening books still place most monocots with six similar "petals" into the lily family. Currently, the traditional lilies are divided into two orders and several other families.

Major changes in the eudicots include the creation of the order Saxifragales. Many of the orders from older plant classifications are no longer recognized or include different families. Some families of eudicots have been combined, while others have been split apart. These changes include:

- The amaranth and goosefoot families have been combined under the name of Amaranthaceae. Chenopodiaceae is no longer recognized as a family.
- The peonies, family Paeoniaceae, are now placed in the order Saxifragales.
- The mallow or hollyhock family, Malvaceae, has several new subfamilies that were formerly independent families. They include the former linden family, Tiliaceae, and the former family Sterculiaceae.
- The maple family, Aceraceae, is no longer recognized on its own. It is now a subfamily of the soapberry family, Sapindaceae.
- The borage family and the waterleaf family are combined; they are called Boraginaceae.
- The milkweed family, Asclepiadaceae, is no longer recognized. It is now a subfamily of the dogbane family, Apocynaceae.
- Most members of the family Scrophulariaceae have been moved to other families, which include Orobanchaceae and Plantaginaceae. Relationships in the former Scrophulariaceae have been difficult to resolve, and their investigation is ongoing.

What you won't see on this tour are the many biochemical and microscopic features that help botanists classify plants. The DNA sequences of selected genes are vital in determining relationships. Features such as how the ovules are attached inside the ovary and the cellular structure of the vascular tissues are examples of some of the additional data used. Bear in mind that you cannot see everything that went into this system of classification. If you would like further details, see Selected References at the book's end.

Before we start: A brief review of the language of botany and flower structure

Foreign language notes: For botanists, words such as "nut" and "numerous" have special meanings, apart from their common usage. In botany, a fruit is not something we eat for dessert. It is a mature, ripened ovary of a flower. It can be hard and dry, or soft and juicy. Botany has many terms, because of the need to describe the huge diversity of plants. As you encounter terms or usages that are foreign to you, consult the glossary for help.

By the way, what IS a flower? A flower is the reproductive organ of a flowering plant. (Note: Pine trees don't have flowers; neither do ferns nor mosses.) Each flower is a puzzle for us to solve, for it has certain basic parts, but they may be greatly modified. We will be seeing and describing many flowers, so you will need some basic terms. I will describe a generalized flower, but bear in mind that there are many variations. The flower starts with its stem, sometimes called a floral shoot. The end of the stem, where the flower attaches, is called the **receptacle**. There are four basic levels of parts attached above this. The first is the **calyx**, which is made up of **sepals**. Sepals usually cover the flower when it is in bud. The next is the **corolla**, which is made up of all the **petals**. The term for the calyx and corolla together is the **perianth**, which literally means "surrounds the flower." If the perianth segments are both colored and look similar, they are neither sepals nor petals, but instead are called **tepals**. Flowers such as tulips and daylilies have tepals, not petals and sepals.

So what is the real flower? That comes next. Above the petals there are **stamens**, the pollen-producing structures, which typically have a stalk-like filament with a pollen-bearing anther at the end. In the center, there are **carpels**, where the seeds develop. There can be a single carpel, several independent ones, or a **pistil** that is made of two or more carpels fused together. Carpels and pistils can have three sections—the **ovary**, the **style**, and the **stigma**. The ovary holds the ovules, the structures that become seeds if they are fertilized. The ovary matures into the fruit of the plant. The style is a connecting structure that ties the ovary to the stigma, the pollen-receiving surface.

These flower parts may not seem exceptionally complicated, but the variations are enormous. Floral parts can be fused either to like or different parts. Parts can be reduced in size, altered to something with a different function, or be missing altogether. The calyx and corolla, which together make up the perianth, can be large and showy, tiny and hard to see, or not there at all. To be a flower, a structure has to have some part that functions in the plant's reproduction. It can be as simple as a single stamen or a single pistil.

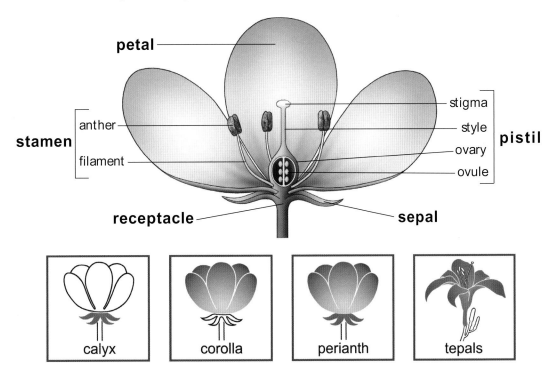

Tree Diagram of Flowering Plant Orders: Nymphaeales

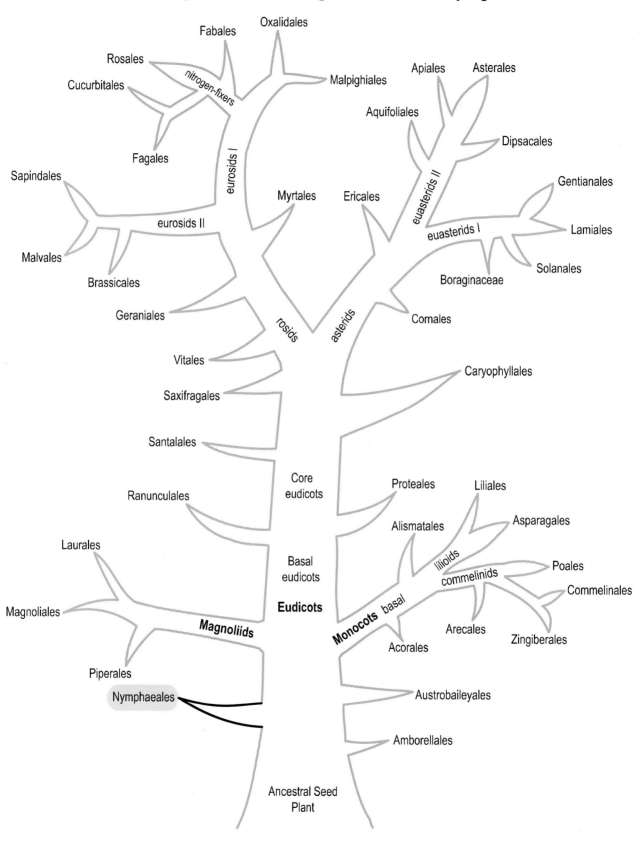

Introducing the First Branches of the Flowering Plants

For many years, botanists have speculated about the oldest lineage of the flowering plants. Magnolias were once thought to be the oldest, based on the fossil record. However, flowering plants do not fossilize very well, probably because they have few hard parts and they grow in environments that lead to decay rather than preservation. With further studies of the "second fossil record," the sequences in DNA, a new picture has come into focus.

Of flowering plants that still exist, the plant from the earliest or most basal branch is *Amborella*, a shrub on the island of New Caledonia. That is outside the realm of this tour, which starts with Nymphaeales, on the second oldest lineage of the presently living flowering plants. Nymphaeales includes the family Nymphaeaceae, the water lilies and pond lilies, and the family Cabombaceae, the fanworts, which are common aquarium plants.

Next to branch was the ancestor of the Austrobaileyales, the order of star anise (*Illicium*), star vine and wild sarsaparilla (*Schizandra*), and Japanese kadsura (*Kadsura)*. Most members of Austrobaileyales are found in Australia and New Guinea or Asia. The North American representatives are two species of star anise and one wild sarsaparilla, which live in the southeastern United States.

After these early branches, the flowering plants split into three main lineages, the magnoliids, the monocots, and the eudicots ("true dicots"). It is difficult to tell if one of these came first or if they all emerged more or less at the same time. Because the magnoliids have traditionally been considered among the oldest lineages, they are often grouped with the basal branches. Together, the basal branches and the magnoliids make up about 3% of flowering plants and are termed the basal angiosperms.

The basal angiosperms were once classified as dicots, but it wasn't a very good fit. They have many characteristics in common with monocots, but do not properly fit on the monocot branch either. Placing them on their own separate branches better reflects flowering plant evolution.

Basal angiosperms

Basal lineages

Nymphaeales Dumortier

    Nymphaeaceae Salisbury   6 genera/58 species
      Genera include:
      Barclaya Wallich—orchid lily, barclaya
      Nuphar Smith—yellow pond lily, spatter dock, beaver-lily, brandy bottle
      *Nymphaea L.[†]—water lily
      *Victoria Lindley—Victorian water lily, Amazon water lily

* illustrated
[†] L. = Carl von Linnaeus

Nymphaeaceae, The Water Lily Family

The family Nymphaeaceae, the water lilies, is a sister group to other flowering plants. Water lilies are aquatic plants whose rhizomes are anchored on the bottom of ponds. Genus *Nymphaea* has flowers with four to six sepals that are sometimes colored like petals. The petals are numerous; they gradually transition into stamens, which are arranged in spirals. These are primitive features. In more highly evolved angiosperms, the flower parts have definite numbers and are arranged in whorls. As in the monocots, grains of water lily pollen have one opening.

Examine the parts near the middle of this water lily flower, a *Nymphaea* hybrid. Some look something like stamens, but also something like petals.

An overview of water lilies (←), *Nymphaea* hybrids, shows their nearly round leaves. The leaves are rolled in bud and unfurl as they mature. The long petioles allow the blades to float on the water's surface. The blades have a deep notch near the petiole attachment.

Victorian or Amazon water lilies have an upturned rim on the edge of their leaves.
(←) *Victoria*, Longwood hybrid

Tree Diagram of Flowering Plant Orders: Magnoliids

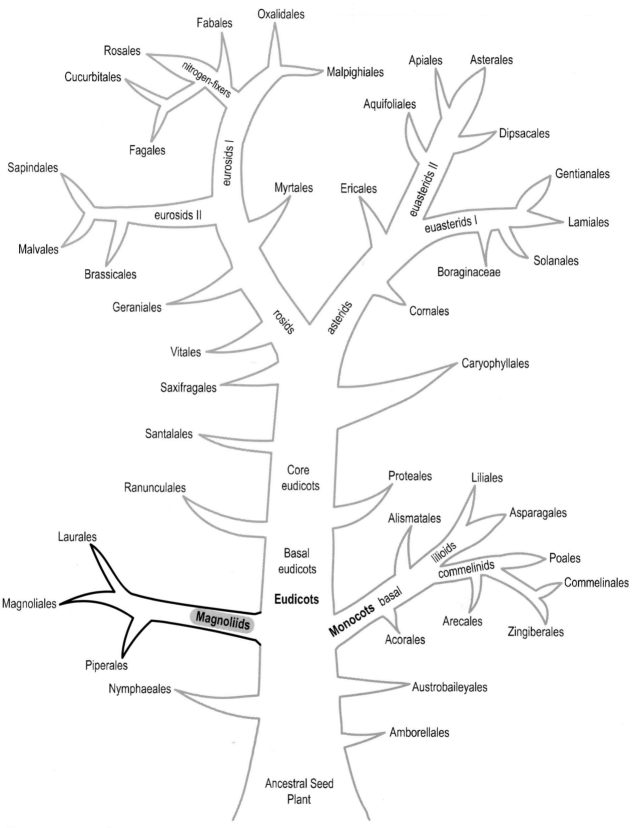

Introducing the Magnoliids

The magnoliid angiosperms were once included with the dicots, but now they are placed on a separate branch of the flowering plants. Many magnoliids have spirally arranged flower parts, more like the spirals of pine cones than the whorls of parts found in monocots and eudicots. Most have pollen with one opening, unlike eudicot pollen, which has three openings.

Four orders make up the magnoliids—Magnoliales, Laurales, Piperales, and Canellales. This tour visits the first three, the magnolia, laurel, and pepper orders. Canellales includes winter's bark, *Drimys*, and white cinnamon, *Canella*, and is found in Central and South America, Australia and New Guinea, and east Africa.

Most magnoliids grow in tropical to mild temperate climates, although some magnolia trees can tolerate colder winters. Several magnoliids, such as the annona or paw-paw family, nutmeg family, and pepper family are characteristic of lowland, moist forests.

There is quite an assortment of flower structure in this group. Some magnoliids have flower parts in multiples of three, as do monocots. The flowers range from the large showy magnolias to the tiny, almost indistinguishable flowers of the pepper family. The leaves have smooth edges, which botanists call entire margins. Magnoliids usually have fragrant oils, and this branch of the flowering plants provides us with the familiar cinnamon, nutmeg, mace, black pepper, and bay leaves.

Basal angiosperms

Magnoliids

Laurales Perleb

Lauraceae Jussieu   50 genera/2500 species

Genera include:

Cinnamomum Schaeffer—cinnamon, camphor tree

  *Cinnamomum verum J. Presl—cinnamon

Laurus L.—bay laurel, sweet bay

Lindera Thunberg—spice bush, wild allspice

  *Lindera benzoin (L.) Blume—spice bush

Litsea Lamark—pond spice

Persea Miller—avocado, red bay tree, silk bay tree

Sassafras Nees & Ebermaier—sassafras

  *Sassafras albidum (Nuttall) Nees—sassafras

Umbellularia (Nees) Nuttall—California bay, Oregon myrtle

* illustrated

Lauraceae, The Laurel Family

(←) Flowers of the spice bush, *Lindera benzoin*, bloom early in the spring. Staminate and pistillate flowers are on separate plants. This shrub grows in woods and bogs of eastern North America. Its fruits are small drupes that turn red when they ripen in the fall (→). (Fruit photo © Irene H. Stuckey, used by permission)

Cinnamomum verum (↓) is one source of the popular culinary spice, cinnamon, which is the inner bark of the tree.

Laurel family members are trees and shrubs that have aromatic oils in their bark and leaves. The flowers have six green, yellow, or white tepals.

The anthers in this family have an unusual way of opening. They have tiny flaps that peel back, pulling out the sticky pollen.

Sassafras trees (←), *Sassafras albidum*, have three different leaf shapes, often on the same branch. The leaves may have three lobes or two lobes in a mitten-like arrangement or no lobes at all. Sassafras is dioecious. The pistillate flowers (→) have yellow staminodes. The dark blue, berry-like fruits are borne in shallow cups at the end of the pedicel (flower stem). The pedicel turns red as the fruit matures, a striking feature.
(Flower photo © Lisa L. Gould, used by permission)

Basal angiosperms

Magnoliids

Laurales

13

Basal angiosperms

Magnoliids

Magnoliales Bromhead

Magnoliaceae Jussieu   2 genera/227 species

Genera include:

Liriodendron L.

*Liriodendron chinense (Hemsley) Sargent—Chinese tulip tree

*Liriodendron tulipifera L.—tulip tree, tulip poplar

Magnolia L.—magnolia, sweet bay, cucumber tree

*Magnolia grandiflora L.—Southern magnolia, bull bay

* illustrated

Magnoliaceae, The Magnolia Family

The primitive features of magnolia flowers include strap-shaped stamens that do not show a distinct filament or anther, and spirally arranged flower parts. The tepals of flowers on one tree can vary from six to 12. Magnolias have many separate carpels. An elongated receptacle holds the spirals of carpels and stamens.

(←, ↓) *Magnolia grandiflora* has white stamens set below green carpels with curled stigmas. The fruit (↓) is an aggregate of follicles. The follicles split on the surface that faces away from the stem, which is unusual, since follicles typically split on the surface that faces the stem. The seeds have a red fleshy coat. They dangle out of the follicles, held by fine fibers. Birds disperse the seed.

(↑) The Chinese tulip tree, *Liriodendron chinense*, and the North American tulip tree (↑), *L. tulipifera*, are the last two survivors of a genus that was once widespread in the Northern Hemisphere. Climate changes in the Tertiary and Quaternary time periods appear to have caused the disappearance of other *Liriodendron* species.

Basal angiosperms

Magnoliids

Piperales Dumortier

Aristolochiaceae Jussieu

4 genera/180 species

Genera include:

Aristolochia L.—Dutchman's pipe, birthwort, pipevine, snakeroot

*Aristolochia clematitis L.—European birthwort

*Aristolochia littoralis Parodi—elegant Dutchman's pipe, calico flower

*Aristolochia watsonii Wooton & Standley—pipevine

Asarum L.—wild ginger, heartleaf

Saruma Oliver

*Saruma henryi Oliver—Chinese wild ginger

* illustrated

Aristolochia flowers are tubular and have a bulbous base. They attract beetles and flies as pollinators. The insects crawl inside the flower, which traps them overnight in the enlarged flower base. The flower releases its pollen and its tube reopens. The pollen-covered insects escape and enter another *Aristolochia* flower.

(←) The flowers of Dutchman's pipe, *Aristolochia littoralis*, attract flies with a dark red face and an odor like rotting meat. The plant is a vine with heart-shaped, or cordate, leaves, which is typical of this family.

The pipevine (→), *Aristolochia watsonii*, is a native to the Sonoran desert. It is a low, spreading vine with arrow-shaped leaves. The flowers are pollinated by tiny, blood-sucking ceratopogonid flies. The musty odor of the flowers and their hairy entrance resemble a rodent's ear, a suitable place for these flies to feed.

(↓) *Aristolochia clematitis* is called European birthwort. It was used supposedly to aid child-birth, but its alkaloids could have poisoned more than they helped.

Saruma henryi (↓) has flowers with three sepals and three petals. Although its common name is Chinese or upright wild ginger, it is not related to culinary ginger, *Zingiber officinale*.

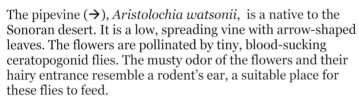

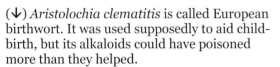

Basal angiosperms

Magnoliids

Piperales

17

Basal angiosperms

Magnoliids

Piperales Dumortier

    Piperaceae C. A. Agardh   6-8 genera/2015 species

Genera include:

Peperomia Ruiz & Pavón—peperomia

    **Peperomia argyreia* C. J. Morren—watermelon peperomia

    **Peperomia caperata* Yuncker—emerald ripple peperomia

*Piper L.—black pepper, white pepper, kava, betel

    *Piper auritum Kunth—Mexican pepperleaf, root beer plant, hoja santa

    Piper nigrum L.—black pepper, white pepper

* illustrated

(←) This peperomia plant, a cultivar of *Peperomia caperata,* is blooming. Its tiny flowers are packed densely on spikes. The flowers have no perianth, so individual ones are hard to see. There are about a thousand species of *Peperomia*, which grow in the tropics worldwide.

Piper is a widespread genus in the tropics; it also has about a thousand species. Its inflorescences are similar to those of *Peperomia*. In Central America, the *Piper* fruits are eaten by bats. The species we use for our black and white pepper seasoning is *Piper nigrum*. It originated in India.

A species of *Piper* from the Costa Rican cloud forest (↓)

(↑) *Peperomia argyreia* is known as watermelon peperomia.

Mexican pepperleaf (←), also called root beer plant, *Piper auritum*, is native to southern Mexico and Central America. It has become naturalized in Florida.

Basal angiosperms

Magnoliids

Piperales

19

Basal angiosperms

Magnoliids

Piperales Dumortier

Saururaceae Martynov   5 genera/6 species

Genera include:

Anemopsis Hooker & Arnott
   *Anemopsis californica (Nuttall) Hooker & Arnott—yerba mansa
Houttuynia Thunberg
   *Houttuynia cordata Thunberg—houttuynia
Saururus L.
   *Saururus cernuus L.—lizard's tail, swamp dragon

* illustrated

Saururaceae, The Lizard-tail Family

(↑, →) The flowers of *Houttuynia cordata* 'Chameleon' have no perianth. They are densely packed on a short spike. There are four white, petal-like bracts at the base of the inflorescence. This colorful ground cover plant originated in east Asia.

Yerba mansa (↓), *Anemopsis californica*, is native to wetlands of the southwestern United States. Its inflorescence has several large white bracts at the base, as well as small white bracts attached to each tiny flower. (Photo © Jay Sullivan, used by permission)

Lizard's tail (→), *Saururus cernuus*, is a native of eastern North America. Its long, tapering inflorescence is covered with tiny white flowers. It grows in moist soil or in shallow water.

Tree Diagram of Flowering Plant Orders: Monocots

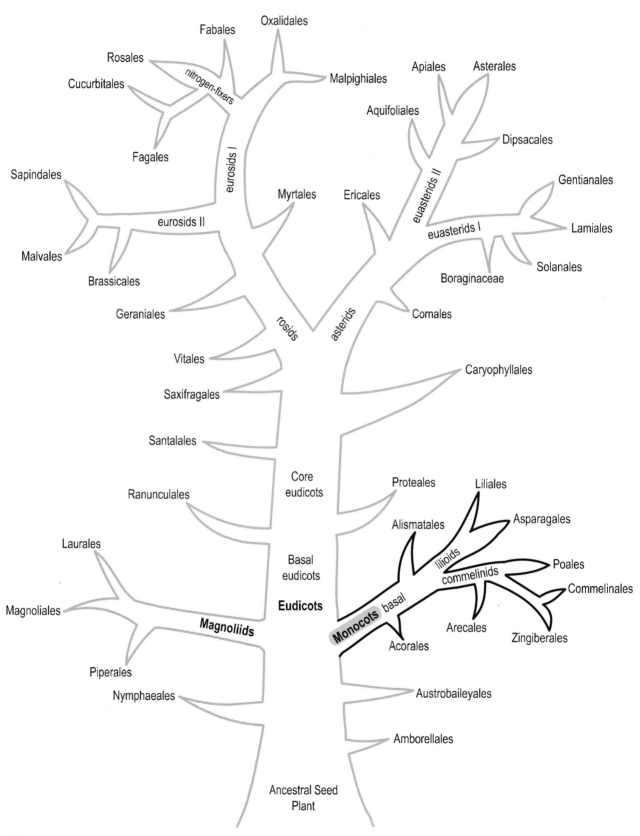

Introducing the Monocots

The monocots have been recognized as a subgroup of flowering plants ever since the seventeenth century. In many traditional classifications, they were one of the two groups of the angiosperms. With recent intensive study of their nucleic acids and all their other characters, the monocots still stand as a single lineage.

That being said, it is not as easy to define monocots as one might think. The textbook description of monocots often states that their leaves have parallel veins. Many do have parallel veins, but there are some with netted venation. Monocot reversion to netted veins is probably an adaptation to dimly lit forest understory environments. It is easy to find these netted vein monocots just by looking in the lobbies of office buildings. They make good indoor plants, since they can live in low light. *Monstera*, *Dieffenbachia*, and *Philodendron* are common examples. To find a monocot with parallel veins, look just about anywhere plants grow for one of the ubiquitous grasses.

Monocot leaves seldom have teeth, but when they do, the teeth have spines and are not glandular-tipped like the teeth of some eudicot leaves. Monocot flowers are usually described as having parts in threes. While this is true, there are also magnoliids whose flowers have parts in threes. The seeds of monocots carry the best clue to their identity; they have a single cotyledon (seed leaf), the trait for which monocots are named. As these seeds sprout, the first root that forms, called the primary root, withers and many adventitious roots grow from the sides of the stem, forming a fibrous root system. Fibrous roots are found in other plant groups, so they are not unique to monocots.

Changes to our understanding of monocots in the Angiosperm Phylogeny Group classification include the establishment of *Acorus*, sweet flag, as the basalmost member and sister group to the rest. The next lineage is Alismatales, the order of water plantains and aroids. Other major branches are the lilioid or petaloid monocots and the commelinid monocots.

The lilioid or petaloid monocots include the lily order, Liliales, and the asparagus order, Asparagales. The traditional lily family was a huge miscellaneous box of unrelated species. Many of its members have been moved to the Asparagales.

The commelinid monocots include the orders of the palms, the grasses, the dayflowers, and the gingers. Of these, the dayflower order, Commelinales, and the ginger order, Zingiberales, are more closely related to each other than they are to the grasses, Poales, and the palms, Arecales.

Monocots

Basal monocots

Acorales Reveal

    Acoraceae Marinov   1 genus/2-4 species

        Sole genus:

        Acorus L.

                *Acorus calamus L.—sweet flag

                Acorus gramineus Solander—Japanese sweet flag

* illustrated

Acoraceae, The Sweet Flag Family

Sweet flag, *Acorus*, is the sole genus of this family. Its common name comes from the aromatic fragrance of its foliage. Its inflorescence is a pale spike, which arises along the side of a leaf. The flowers have their parts in threes, although they are so tiny that this is hard to see. Sweet flag has flat leaves, something like those of the iris, but with their midrib set off center. This family is considered to be the first branch of the monocots and a sister group to the rest.

(↑) The young spikes of sweet flag are green. They appear cream-colored as the flowers mature (↑). The spike is covered in the tiny flowers, which have six tepals and six stamens. The fruits are berries.

An overview of sweet flag (↓)

Fruits are developing on this spike (↓).

Monocots

Basal monocots

Alismatales Dumortier

Alismataceae Ventenat   12 genera/81 species

Genera include:

Alisma L.—water plantain

Damasonium Miller—star water plantain

Echinodorus Richard ex Engelmann—aquarium swordplant, burrhead

Sagittaria L.—arrowhead, wapato

Sagittaria lancifolia L.—bulltongue arrowhead, duck potato

* illustrated

Alismataceae, The Water Plantain Family

The flowers of this family have three sepals and three petals. *Alisma* has bisexual flowers. *Sagittaria* has unisexual flowers. The many carpels are separate rather than fused together. The fruits are achenes that float. They are dispersed on water and by waterfowl that eat them.

(↑) *Alisma*, commonly called water plantain, has bisexual flowers (↑) in an inflorescence with many whorled branches. Its flowers are smaller than those of *Sagittaria* (below).

Arrowhead, *Sagittaria,* has staminate and pistillate flowers, usually on the same plant. The flowers form in whorls of three; these whorls are the only branches of the inflorescence. Leaf shape varies a great deal within the genus and even within a species, depending on growing conditions. Some species even have grass-like leaves. The starchy edible tubers of some species give them the common name of duck potato.

This *Sagittaria* (↓) has the typical leaf shape from which the common name, arrowhead, is derived. The flowers are staminate.

Sagittaria lancifolia has ovate leaves. It is shown with staminate flowers above the developing fruits (↓).

Monocots

Basal monocots

Alismatales Dumortier

Araceae Jussieu   106 genera/4025 species

Genera include:

Aglaonema Schott—Chinese evergreen

Alocasia (Schott) G. Don—elephant's ear, taro

Anthurium Schott—anthurium

Arisaema Martius—arisaema, green dragon, cobra lily, jack-in-the-pulpit

Caladium Ventenat—caladium

    *Caladium bicolor (Aiton) Ventenat—fancy-leaf caladium

Colocasia Schott—taro

*Dieffenbachia Schott—dieffenbachia, dumb cane

Epipremnum Schott—pothos, tongavine

*Lemna L.—duckweed

Lysichiton Schott—skunk cabbage

Monstera Adanson—swiss cheese plant, split leaf philodendron

Orontium L.—golden club

Peltandra Rafinesque—arrow arum

Philodendron Schott—philodendron

Pistia L.—water lettuce

Scindapsus Schott—satin pothos

*Spathiphyllum Schott—peace lily

Symplocarpus Salisbury ex Nuttall

    **Symplocarpus foetidus* (L.) Salisbury ex Nuttall—skunk cabbage

Syngonium Schott—arrowhead vine

Wolffia Horkel ex Schleiden—water meal (the smallest flowering plant)

Xanthosoma Schott—tannia, yautia

Zantedeschia Sprengel—calla lily

    **Zantedeschia aethiopica* (L.) Sprengel—white calla lily

* illustrated

Araceae, The Aroids or Arum Family

Peace lily (↑), *Spathiphyllum*

The tiny flowers of aroids are embedded on a spike that is called a spadix. The spadix is surrounded by a large bract, the spathe, which is often brightly colored. (↓) In the calla lily, the lower spadix flowers are pistillate and the upper ones are staminate. Some aroids have bisexual flowers. The leaf shape and venation vary tremendously in this family.

Dieffenbachia in bloom (→)

(↓) *Symplocarpus foetidus* is the skunk cabbage of eastern North America.

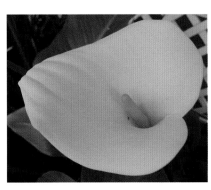

The calla lily, *Zantedeschia aethiopica*, with an overview of its white spathe (↑) and close-up of its golden spadix (→).

Caladium bicolor (↓) has beautifully patterned leaves, which are good examples of netted veins in a monocot.

The tiny duckweeds (↓), genus *Lemna*, have minute flowers with single stamens and carpels, but these aquatic plants typically reproduce vegetatively.

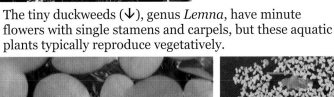

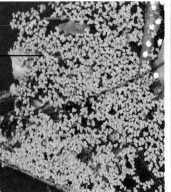

Monocots

Lilioid (petaloid) monocots

Asparagales Bromhead

Agavaceae Dumortier   23 genera/637 species

Genera include:

*Agave L.—century plant, lechuguilla, mescal

*Camassia Lindley—camass, quamash

Chlorogalum Kunth—soap plant, amole

Hesperaloe Engelmann—red yucca

Hesperocallis A. Gray—desert lily, ajo lily

*Hosta Trattinick—hosta, plantain lily

*Yucca L.—yucca

   *Yucca rigida Trelease—blue yucca, silver yucca

* illustrated

The Breakup of the Traditional Lily Family

Previously, most monocots with six tepals and six stamens were placed in the lily family. With more information about all the details of plant structure and development, the former lily family was divided between two orders, Asparagales and Liliales. Seeds of Asparagales usually have a black outer layer not found in the Liliales. Asparagales flowers have nectaries on the side of the ovary and tepals that usually lack spots. Liliales flowers have nectaries at the base of their tepals and are often spotted. Studies of their DNA show that the two orders are quite different branches of monocots. Field guides and gardening books will probably still list many of the Asparagales families as part of traditional Liliaceae.

Agavaceae, The Agave Family

Century plant (↑), *Agave*

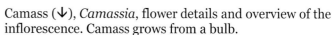

Plantain lily (↑), *Hosta* hybrid, overview of plant and close-up of flowers

The agave family includes many plants that are adapted to live in arid environments, as well as some adapted to moist, shady woodlands. The plants have a basal rosette of leaves. Most grow from rhizomes, but a few form bulbs. The seeds are flat, with a black outer layer. Yuccas have a special pollination partnership with yucca moths. Their coevolution is a classic example of the symbiotic relationship, mutualism.

Camass (↓), *Camassia*, flower details and overview of the inflorescence. Camass grows from a bulb.

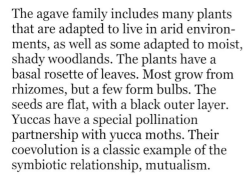

Yuccas bloom many times during their lifetime, but they may not bloom every year, depending on the moisture and other resources available. Most agaves bloom only once. The basal rosette of leaves accumulates food for years, then blooms and dies. Some agaves are called century plants, but it usually takes just a few decades for them to build enough resources to bloom.

(↑) Blue yucca, *Yucca rigida*, whole plant and flower

The dried, three-part capsule of a yucca (↑)

Monocots

Lilioid (petaloid) monocots

Asparagales Bromhead

      Alliaceae Batsch ex Borkhausen   13 genera/795 species

          Genera include:

          Allium L.—onion, leek, garlic, shallot, allium

               *Allium cernuum Roth—nodding onion

               *Allium giganteum Regel—giant allium

               *Allium schoenoprasum L.—chives

          Ipheion Rafinesque

               *Ipheion uniflorum (Graham) Rafinesque—spring star flower

          Nothoscordum Kunth—false garlic, crow poison

          Tulbaghia L.—society garlic

* illustrated

Alliaceae, The Onion Family

The genus *Allium* includes giant allium (↑, ↓), *A. giganteum*, and chives (←), *A. schoenoprasum*.

The fruits of giant allium (↑) split into three sections when they are mature, revealing the black seeds.

The onion family has sulfur-containing molecules that irritate people's eyes. The flowers have six tepals and grow in umbels with scaly bracts at their base. They have a superior ovary. These herbaceous plants form bulbs, structures composed of leaves that are modified for storage. Some species, such as chives and onions, have tubular leaves. Seeds may be triangular, tetrahedral, or rounded.

The nodding onion (↓), *Allium cernuum*, is a native wildflower of wooded areas in much of the United States and southern Canada.

Spring star flower, *Ipheion uniflorum* (→), has three yellow stamens that are clearly visible and three that attach further down in the narrow tube formed by the fused bases of its six tepals.

Monocots

Lilioid (petaloid) monocots

Asparagales Bromhead

Amaryllidaceae J. Saint-Hilaire   59 genera/800 species

Genera include:

Amaryllis L.—naked lady, Cape belladonna, belladonna lily

Clivia Lindley—clivia

       *Clivia miniata (Lindley) Regel—clivia, kaffir lily, bush lily

Crinum L.—swamp lily, St. John's lily, creole lily, crinum

Galanthus L.—snowdrop

       *Galanthus nivalis L.—common snowdrop

*Hippeastrum Herbert—amaryllis

Hymenocallis Salisbury—spider lily, Peruvian daffodil

Leucojum L.—snowflake, snowdrop

*Narcissus L.—daffodil, narcissus, jonquil, paper white

Sprekelia Heister—Jacobean lily, Aztec lily

Zephyranthes Herbert—zephyr lily, atamasco, rain lily

* illustrated

Amaryllidaceae, The Amaryllis or Daffodil Family

Daffodil (←),
Narcissus hybrid

Kaffir lily or clivia
(→), *Clivia
miniata*

Snowdrops (↓),
Galanthus nivalis

The flowers of Amaryllidaceae have inferior ovaries (←, white arrows). Daffodils have a corona, an extension of the perianth. The flowers may be solitary or in few-flowered, umbel-like inflorescences. The plants of this family grow from bulbs.

The amaryllis (↑), a hybrid of genus *Hippeastrum*, grows from a large bulb. It is often grown indoors in the winter. The bloom stalk is usually a foot or more tall. Several flowers form at its end, with their buds initially enclosed in a large, green bract. When the tepals first open, the stamens and pistil are immature. (↑) The stamens mature first. Their anthers start as long, pale lavender or cream-colored structures, which split and fold back to reveal the yellow, pollen-bearing surface. The style grows longer, but the stigma is still closed when the anthers first split (↑). Finally, the style bends upward (↑) and the three lobes of the stigma open, readying the flower to receive pollen.

Monocots

Lilioid (petaloid) monocots

Asparagales Bromhead

  Asparagaceae Jussieu   2 genera/165-295 species

    Genera are:

    Asparagus L.—asparagus, asparagus fern

      *Asparagus densiflorus (Kunth) Jessop—Sprenger's asparagus fern

      *Asparagus officinalis L.—garden asparagus

      *Asparagus setaceus (Kunth) Jessop—asparagus fern

    Hemiphylacus S. Watson

* illustrated

Asparagaceae, The Asparagus Family

The genus *Asparagus* holds all but a few species of this family. Many Asparagus species have green, needle-like branches that perform most of the plant's photosynthesis. The leaves are reduced to tiny scales. The flowers are small, with six greenish-yellow or white tepals. The fruit is a berry that may be red, blue, or black.

Garden asparagus (→), *Asparagus officinalis*

(↑) *A. densiflorus* Sprengeri is often grown as an ornamental. (↑) It has small, white flowers and (↑) long, arching branches with narrow leaves. The fruits are red berries.

The fruits of garden asparagus are berries (↑) that have a persistent perianth.

The triangular structures (→) on a new asparagus shoot are derived from true leaves.

(↑) *Asparagus setaceus* is called asparagus fern, but it is not a spore-bearing plant. It is a true *Asparagus* as its new shoots show; they do not resemble fern fiddleheads. Furthermore, it bears tiny white flowers and blue-black berries, proof that it is a flowering plant.

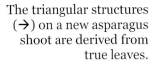

Monocots

Lilioid (Petaloid) Monocots

Asparagales

37

Monocots

Lilioid (petaloid) monocots

Asparagales

Hemerocallidaceae R. Brown   19 genera/85 species

Genera include:

Dianella Lamark ex Jussieu—flax lily

**Hemerocallis* L.—daylily

Phormium Forster & G. Forster

  **Phormium tenax* Forster & G. Forster—New Zealand flax, harakeke

* illustrated

Hemerocallidaceae, The Daylily Family

Daylilies (←), *Hemerocallis* hybrids, are popular garden flowers. They have been bred in many colors, including yellow, orange, pink, and red. Their foliage is a basal clump of long, narrow leaves. Their flowers last only one day, hence their name. Daylilies were introduced to North America from Europe. Some varieties have escaped cultivation and become invasive in parts of the eastern United States.

(←) The style of the daylily is longer than the stamens. At the tip there is a small stigma (white arrow), about the same diameter as the style.

The base of the flower is a floral tube (↑, white arrow), a short cylinder that is formed from the fusion of the six tepals. The ovary, which is superior, sits inside the floral tube.

(←) New Zealand flax, *Phormium tenax*, is not related to the European flax, *Linum*, that has been used to make linen cloth since prehistoric times. The Maori, the first settlers of New Zealand, used the fibers from *Phormium* leaves to make clothing and cordage. Europeans called the plant New Zealand flax because it was a good source of fiber. Horticulturalists have bred variegated and colored-leaf varieties of *Phormium* for landscape use around the world.

Monocots

Lilioid (petaloid) monocots

Asparagales Bromhead

Hyacinthaceae Borkhausen · 41-70 genera/1000 species

Genera include:

Brimeura Salisbury—brimeura

Chionodoxa Boissier—chionodoxa

> **Chionodoxa luciliae* Boissier—glory-of-the-snow

Eucomis L'Héritier—pineapple lily, eucomis

Galtonia Decaisne—summer hyacinth

Hyacinthoides Heister ex Fabricius—wood hyacinth, English and Spanish bluebell

Hyacinthus L.—hyacinth

> **Hyacinthus orientalis* L.—garden hyacinth

Muscari Miller—grape hyacinth

> **Muscari botryoides* (L.) Miller—common grape hyacinth

Ornithogalum L.—star of Bethlehem, pregnant onion

Puschkinia Adams—puschkinia

> **Puschkinia scilloides* Adams—striped squill

Scilla L.—squill

* illustrated

Hyacinthaceae, The Hyacinth Family

Striped squill, (↑) *Puschkinia scilloides*

Hyacinths (↑, ↗), *Hyacinthus orientalis*

These spring-blooming flowers grow from bulbs. The leaves grow in a basal rosette from which the flower stalk arises. The flowers are borne on a scape, a leafless stalk. The three-carpellate pistil has a single style. The family members pictured above all originated in the Middle East. The hyacinth family also has many members in Africa. One species of wood hyacinth, in the genus *Hyacinthoides*, is native to North America.

Glory-of-the-snow (↓), *Chionodoxa luciliae*

Grape hyacinth (↓), *Muscari botryoides*

Monocots

Lilioid (petaloid) monocots

Asparagales Bromhead

Iridaceae Jussieu   67 genera/1800 species

Genera include:

Belamcanda Adanson—blackberry lily

Crocosmia Planchon—crocosmia

*Crocus L.—crocus, saffron crocus

Dierama K. Koch—fairy wand

Freesia Ecklon ex Klatt—freesia

*Gladiolus L.—gladiolus

*Iris L.—iris

       *Iris missouriensis Nuttall—wild iris

Ixia L.—African corn lily

Sisyrinchium L.—blue-eyed grass

Sparaxis Ker Gawler—harlequin flower

Tigridia Jussieu

      *Tigridia pavonia (L. f.†) DC.‡—tiger flower, Mexican shell flower

* illustrated

† L. f. = Carl von Linneaus, fils

‡ DC. = Augustin de Candolle

42

Iridaceae, The Iris Family

Iris missouriensis (←, →) is native in much of the western United States.

The genus *Iris* has a unique flower structure. Generally, there are three tepals that hang down and three that point up. The three-carpellate pistil has three styles. The ovary is inferior (→). The styles look like petals (←, white arrow). Like all members of the family, this flower has three stamens. They are located under the three petal-like styles. The stigma is a little flap on the underside of the style.

Many iris species and garden hybrids grow from rhizomes.

Gladiolus (←) and *Crocus* (→) grow from corms.

Tiger flower (↓), *Tigridia pavonia*, grows from bulbs.

In Greek mythology, Iris was the goddess of the rainbow. The beautiful colors of hybrid bearded irises (↓) make *Iris* an appropriate name for this genus.

Monocots

Lilioid (petaloid) monocots

Asparagales Bromhead

Orchidaceae Jussieu   788 genera/20,000 species

Genera native to temperate North America include:

Calypso Salisbury

       *Calypso bulbosa (L.) Oakes—fairy slipper orchid

Corallorhiza Gagnebin—coral root orchid

Cypripedium L.—lady slipper orchid

Goodyera R. Brown—rattlesnake plantain

Limnorchis Rydberg—bog orchid

Listera R. Brown—twayblade

Platanthera Richard—fringed orchid, green orchid, bog orchid

       *Platanthera dilatata (Pursh) Lindley ex Beck—tall white bog orchid, scentbottle

Spiranthes Richard—ladies' tresses

Other genera include:

Blettilla Reichenbach f.—urn orchid, ground orchid, blettilla

Cattleya Lindley—cattleya

Dendrobium Swartz—dendrobium

Epidendrum L.—star orchid

       *Epidendrum radicans Pavón ex Lindley—crucifix orchid

*Miltonia Lindley—pansy orchid

*Paphiopedilum Pfitzer—slipper orchid

*Phalaenopsis Blume—moth orchid

Vanda W. Jones ex R. Brown—vanda, lei orchid

Vanilla Plumier ex Miller—vanilla orchid

* illustrated

44

Orchidaceae, The Orchid Family

The orchid family is one of the largest families of flowering plants. Its nearly 20,000 members are mostly tropical. Orchids have tiny, dust-like seeds that must have the help of fungi to germinate. The flowers have six perianth segments, one of which is often modified into a lip or pouch. Orchids usually have one or two stamens, which are fused to the stigma and style. Together, they form a structure called the column.

Fairy slipper (←), *Calypso bulbosa*, and tall white bog orchid (→), *Platanthera dilatata*, are native to cold winter climates of Canada and the western and northern United States.

Most native orchids have small flowers that are hard to see. The hybrids and florists' orchids are more often showy, large-flowered varieties.

Epidendrum radicans (→) has become a weed in Australia and Florida. It is native to Mexico and Central America.

This *Phalaenopsis* hybrid (↑) is cultivated commercially.

Miltonia hybrids (↓) are called pansy orchids.

Many orchids are easily hybridized. Different genera can be crossed, and it can be difficult to determine the parentage of hybrids.

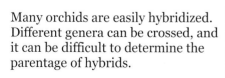

Paphiopedilum hybrid (→)

Monocots

Lilioid (petaloid) monocots

Asparagales Bromhead

Ruscaceae Sprengel   26 genera/475 species

Genera include:

Convallaria L.—lily-of-the-valley

Dracaena Vandelli ex L.—lucky bamboo, corn plant, dracaena
> **Dracaena draco* (L.) L.—dragon tree

Maianthemum F. H. Wiggers—false Solomon's-seal, Canada mayflower
> **Maianthemum stellatum* (L.) Link—false lily-of-the-valley,
> star Solomon's-seal

Nolina Michaux
> **Nolina microcarpa* S. Watson—beargrass, sacahuista

Polygonatum Miller—Solomon's seal

Ruscus L.
> **Ruscus hypoglossum* L.—butcher's broom

Sansevieria Thunberg
> **Sansevieria trifasciata* Prain—snake plant, mother-in-law's tongue,
> bowstring hemp

* illustrated

Related Families with Notable Characteristics

This tour moves through the families in each order alphabetically. The alphabetical arrangement was necessary, given our incomplete knowledge of all the family branching, but it hides some interesting relationships. Certain families within the Asparagales are more closely related than others. These relatives share interesting characteristics.

Asparagaceae and Ruscaceae are closely related. Members from both families have phylloclades, also called cladophylls, which are stems that modified in form and perform the photosynthesis that leaves usually do. The short, narrow phylloclades of asparagus don't look a great deal like the broad ones of *Ruscus*, but they are derived from the same plant part and do the same job.

Ruscaceae is also closely related to a Southern Hemisphere family, Laxmanniaceae. These two, along with the Agavaceae, have family members that form soft wood and grow as trees. Wood is rare in monocots and is formed differently from the much more common wood of eudicots or magnoliids.

Ruscaceae, The Butcher's Broom Family

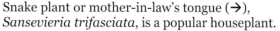

Star Solomon's-seal (←), *Maianthemum stellatum*, is a native to much of the United States, except for the southeast.

Snake plant or mother-in-law's tongue (→), *Sansevieria trifasciata*, is a popular houseplant.

The dragon tree (↑), *Dracaena draco*, is one of the few monocots that form woody stems. Its sap, which is used as a varnish, is deep red and is known as dragon's blood.

Butcher's broom (↑), *Ruscus hypoglossum*, is a remarkably deceptive plant. It appears to have small flowers growing out of the middle of its leaves. The trick is that these are not leaves. They are broad, flattened stems, known botanically as phylloclades. After the butcher's broom blooms, the fruit, which is a red berry, forms in the middle of the phylloclade. This *Ruscus*, like *Nolina*, has unisexual flowers.

Nolina microcarpa (↓), commonly known as beargrass, has narrow leaves that are often frayed at the tips. Its staminate flowers (→) and pistillate flowers are borne on separate plants.

Monocots

Lilioid (petaloid) monocots

Liliales Perleb

Alstroemeriaceae Dumortier — 3 genera/165 species

Genera include:
Alstroemeria L.—alstroemeria, Chilean lily, Peruvian lily
Bomarea Mirbel—climbing alstroemeria, bomarea

* illustrated

Alstroemeriaceae, The Alstroemeria Family

This family grows in Central and South America. *Alstroemeria* is known as the Peruvian or Chilean lily. *Alstroemeria* hybrids are florist's favorites, valued for their long-lasting cut flowers.

(↑) When the *Alstroemeria* flower first opens, the anthers split and release their yellow pollen, but the stigma is still closed.

In older flowers, the three branches of the stigma open (↑, ↗), and the anthers may fall off.

The ovary is inferior (↑, white arrow) in this family. After five of the six tepals and the stamens are removed, the pistil is clearly visible (↑), revealing its inferior ovary, long style, and three stigma branches. Note the twisted petiole of the leaves, a characteristic of this family.

Monocots
Lilioid (petaloid) monocots
Liliales Perleb

Liliaceae Jussieu   16 genera/635 species

Genera include:

Calochortus Pursh—sego lily, mariposa lily, star tulip
       **Calochortus gunnisonii* S. Watson—Gunnison's mariposa lily
Clintonia Rafinesque—bluebeard, yellow bead lily
Erythronium L.—fawn lily, trout lily, dogtooth violet
Fritillaria L.—fritillary, checker lily, chocolate lily, snakeshead
       **Fritillaria imperialis* L.—crown imperial
*Lilium L.—lily, tiger lily, Easter lily, Turk's cap lily
Streptopus Michaux—twisted stalk
Tricyrtis Wallich—toad lily
       **Tricyrtis formosana* Baker—Formosa toad lily
*Tulipa L.—tulip

* illustrated

Liliaceae, The Lily Family

Asian lily (↑), *Lilium* hybrid

Mariposa or sego lily (↑),
Calochortus gunnisonii

Crown imperial (↑),
Fritillaria imperialis

Once the lily family held nearly all the lilioid monocots with six tepals. When botanists studied the plants further, they moved many species to other families. True lilies usually grow from bulbs with contractile roots. Their nectaries are located at the base of their tepals. Many lilies have spotted tepals, and their tepals are distinct, not fused. The ovary is superior (arrow, →). The fruit is usually a capsule, but may also be a berry. Lily pollen is large and can be seen with only a little magnification.

Lilium hybrid (→)

The toad lily (←), *Tricyrtis formosana*, is native to Taiwan. The toad lilies have speckled flowers with three extended, down-curving stigma lobes, each of which branches in two at the end. The filaments of the stamens grow close to the ovary.

The tulip (↓), a *Tulipa* hybrid, has a superior ovary (↓) that develops into a three-part capsule (↓).

Monocots

Commelinid monocots

Arecales Bromhead

Arecaceae Schultz-Schultzenstein or Palmae Jussieu

190 genera/2500 species

Genera include:

Calamus L.—rattan palm, lawyer cane
*Caryota L.—fishtail palm
*Cocos L.—coconut palm
Copernicia Martius ex Endlicher—carnuba wax palm
Kerriodoxa Dransfield
      *Kerriodoxa elegans Dransfield—white elephant palm, King Thai palm
*Phoenix L.—date palm, phoenix palm
Raphia P. Beauvois—raffia palm
Sabal Adanson—palmetto palm
Washingtonia H. Wendland—California fan palm, Mexican fan palm

* illustrated

Arecaceae or Palmae, The Palm Family

The palm family is the sole family of its order. Palms are trees and shrubs of tropical and subtropical climates. They form a large family of about 2500 species. Palm leaves may be fan-folded, and may be simple or compound. The petiole base has a large fibrous sheath. Palms usually have an unbranched trunk.

(←) *Cocos,* coconut palm

(↓) *Phoenix*, date palm

Kerriodoxa elegans (↑) has beautiful fan-shaped leaves. This ornamental palm is native to Thailand.

The fishtail palm (↓), *Caryota* species, has inflorescences with many hanging branches that resemble a mop.

Palms may have bisexual or unisexual flowers. They may be pollinated either by wind or insects.
Staminate flowers (↗)
Pistillate flowers (↘)

Monocots

Commelinid monocots

Commelinales Dumortier

Commelinaceae Mirbel   40 genera/652 species

Genera include:

Callisia Loefling—inch plant, basket plant

Commelina L.—dayflower

> *Commelina erecta* L.—erect dayflower, slender dayflower, whitemouth dayflower

Tradescantia L.—spiderwort, Moses-in-the-cradle, inch plant

> *Tradescantia occidentalis* (Britton) Smyth—western spiderwort, prairie spiderwort
>
> *Tradescantia pallida* (Rose) D. R. Hunt—purple heart, purple queen
>
> *Tradescantia zebrina* Loudon—wandering Jew

* illustrated

Western or prairie spiderwort, *Tradescantia occidentalis* (↗), is a native of the North American midcontinent. Horticultural varieties of this species (↑) are grown as ornamentals.

The spiderwort family has sheathing leaves and flower clusters at the ends of the stems. The stamens often have fringed filaments (↑). The flowers have a calyx of three sepals and a corolla of three petals, one of which may be much smaller than the others. The flowers last for only one day.

There are two large, leafy bracts beneath the flowers of this purple heart (←), *Tradescantia pallida*, an arrangement that is common in this family.

Genus *Commelina* is commonly known as the dayflower. The flowers of *C. erecta* (↑) have two large blue petals and one much smaller white one. The stamens have different shapes and functions. Three of them produce pollen normally. Three have little pollen, but their large yellow anthers apparently attract insects.

This wandering Jew (←), *Tradescantia zebrina*, is one of several varieties with different leaf markings.

Monocots

Commelinid monocots

Commelinales Dumortier

Pontederiaceae Kunth   9 genera/33 species

Genera include:

Eichhornia Kunth—water hyacinth

   Eichhornia crassipes (Martius) Solms—common water hyacinth

Heteranthera Ruiz & Pavón—mudplantain

Pontederia L.

   Pontederia cordata L.—pickerel weed

* illustrated

Common water hyacinth (↑), *Eichhornia crassipes,* can be grown in water gardens as an ornamental. This plant originated in tropical South America, but now it has become a pesky weed in warm, slow-moving freshwater in North America (↑), Asia, Africa, and Australia. The leaves have petioles that are enlarged, hollow, air-filled tubes. These keep the plant afloat.

Pickerel weed (↓), *Pontederia cordata,* is a North American native that has an amazing system of breeding. The flowers can have long, medium, or short styles. They have stamens that are different lengths than the styles, either longer or shorter or both. Pollen from any plant is less likely to reach its own stigmas, which promotes genetic diversity.

Monocots

Commelinids

Commelinales

57

Monocots

Commelinid monocots

Poales Small

Bromeliaceae Jussieu   57 genera/1400 species

Genera include:

Aechmea Ruiz & Pavón
> **Aechmea fasciata* (Lindley) Baker—silver vase bromeliad

Alcantarea (E. Morrin ex Mez) Harms
> **Alcantarea imperialis* (Carrière) Harms—imperial bromeliad

Ananas Miller—pineapple
> **Ananas comosus* var. *variegatus* (Lowe) Moldenke—variegated pineapple

Guzmannia Ruiz & Pavón—guzmannia

Neoregelia L. B. Smith
> **Neoregelia carolinae* (Beer) L. B. Smith—heart of flame, blushing bromeliad

Tillandsia L.—tillandsia
> **Tillandsia cyanea* Linden ex K. Koch—pink quill
> **Tillandsia usneoides* (L.) L.—Spanish moss, old man's beard

Vriesea Lindley—lobster claw, flaming sword

* illustrated

58

Bromeliaceae, The Bromeliad Family

Ornamental pineapple (→), *Ananas comosus* var. *variegatus*

Alcantarea imperialis (←) shows the typical central cup formed from overlapping leaf bases.

Members of the bromeliad family have short or sessile stems. Their leaves form a rosette, which holds a water supply in many species. Most bromeliads are epiphytic. The leaves of the epiphytes have scale-like structures that help them absorb this water. The flowers usually have a bract at their base. In many species, the bracts are usually larger and more colorful than the flower itself.

Aechmea fasciata (↓) has pink bracts with tiny purple flowers among them.

The flowers of heart-of-flame bromeliad (↑), *Neoregelia carolinae,* form in the central cup. The leaves that surround the cup develop bright color as the plant blooms.

Unlike those of most bromeliads, flowers of pink quill bromeliad, *Tillandsia cyanea,* have three large, showy petals (↑). They protrude from the edge of the pink bracts. Tillandsias include Spanish moss (←), *T. usneoides,* and other epiphytes known as air plants and ball moss.

Monocots

Commelinid monocots

Poales Small

Cyperaceae Jussieu   98 genera/4350 species

Genera include:

Carex L.—sedge
> **Carex aquatilis* Wahlenberg—water sedge
> **Carex grayi* Carey—Gray's sedge
> **Carex morrowii* Boott—Japanese sedge

Cyperus L.—flatsedge, papyrus, nut grass, tiger nut, chufa
> **Cyperus alternifolius* L.—umbrella plant
> *Cyperus papyrus* L.—papyrus

Eleocharis R. Brown—spikerush, Chinese water chestnut
Eriophorum L.—cottonsedge, bogwool, cotton-grass
Kobresia Willdenow—bogsedge
Schoenplectus Palla—tule, bulrush
> **Schoenoplectus tabernaemontani* (C. C. Gmelin) Palla—
>   soft-stemmed bulrush, great bulrush

Scirpus L.—bulrush, woolgrass

* illustrated

60

Cyperaceae, The Sedge Family

"Sedges have edges..." as the saying goes. Many sedges have triangular stems with edges that are easily felt and may even cut. The edges are a good way to distinguish sedges from grasses, which they resemble. While most sedges prefer aquatic habitats, some species grow away from water. Their flowers have no visible perianth; they are borne in spike-like inflorescences of bracts.

Water sedge (←), *Carex aquatilis*, is a member of a large, monoecious genus. The cinnamon-brown staminate flowers are on the top. The pistillate flowers have white stigmas that grow from black bracts. Members of this genus commonly grow in and near water and on dry land as well.

Umbrella plant (↑), *Cyperus alternifolius*, is used as an ornamental in water gardening. Its pistillate flowers have curled, white stigmas (↑). The inflorescences are flattened spikes. Another member of this genus, *C. papyrus*, is the plant that was used by the ancient Egyptians to make paper.

Gray's sedge (↓), *Carex grayi*, is native to the eastern United States. Its fruits are enclosed by inflated bracts, so that each pistillate inflorescence looks like a spiky ball.
Variegated forms of Japanese sedge (↓), *Carex morrowii*, are grown as ornamentals.

The bulrushes are common aquatic plants of the sedge family. (↓) The soft-stemmed bulrush, *Schoenoplectus tabernaemontani*, has round stems instead of the triangular ones typical of sedges. Its leaves are reduced to small bracts at the base of the stem. Its flowers are bisexual.

staminate flowers

pistillate flowers

Monocots

Commelinid monocots

Poales Small

Poaceae (R. Brown) Barnhart or Gramineae Jussieu

650 genera/9700 species

Genera of cereal grains include:
Avena L.—oats
　　　Avena fatua L.—wild oats
Hordeum L.—barley
　　　Hordeum jubatum L.—foxtail barley
Oryza L.—rice
Secale L.—rye
Sorghum Moench—sorghum, milo
Triticum L.—wheat, spelt, kamut
Zea L.—corn (known as maize in Europe)
Zizania L.—wild rice

Other Poaceae genera include:
Andropogon L.—bluestem grass
　　　Andropogon gerardii Vitman—big bluestem
Bambusa Schreber—bamboo
　　　Bambusa textilis McClure—weaver's bamboo
Bouteloua Lagasca—grama grass
　　　Bouteloua gracilis (Kunth) Griffiths—blue grama grass
Briza L.—quaking grass
Bromus L.—brome grass, cheat grass
Elymus L.—wild rye, wheatgrass
Festuca L.—fescue grass
Panicum L.—panic grass, witch grass, millet
　　　Panicum miliaceum L.—white millet, proso millet
Pascopyrum A. Löve
　　　Pascopyrum smithii (Rydberg) A. Löve—western wheatgrass
Phragmites Adanson—reed, giant reed
Phleum L.—timothy grass
Phyllostachys Siebold & Zuccarini—hardy bamboo
Poa L.—bluegrass

* illustrated

Poaceae or Gramineae, The Grass Family

The grasses are a very large and diverse family. Their leaves are sheathed and most species have hollow stems. It takes careful observation and a hand lens to see grass flowers in bloom. They have no obvious perianth and are typically wind-pollinated. Flowers form within several green bracts that make up a spikelet. When the flower blooms, the bracts are pushed apart. The stamens hang out where the wind can blow away their pollen. The stigmas have many finely divided branches, so they look like tiny feathers. After the flowers bloom, the bracts of the spikelet close back together and the grass's fruit or grain, grows inside.

Big bluestem grass (←), *Andropogon gerardii*, has inflorescences that are typically in groups of three. This plant has pale pink, feathery stigmas and red-brown stamens.

This western wheatgrass (→), *Pascopyrum smithii*, has yellow stamens and small, white stigmas (arrow).

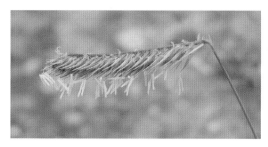

Blue grama grass (←), *Bouteloua gracilis*, is an important component of short grass prairies.

Like the oats we eat, wild oats (↓), *Avena fatua*, belong to genus *Avena*.

Bamboos make up one of several subfamilies of Poaceae. There are over 1000 species. Flowering in the woody bamboos is remarkable, as it occurs only at long intervals, as much as 120 years. The whole bamboo grove flowers at once, and then dies. A new stand grows from the seeds.

Weaver's bamboo (→), *Bambusa textilis*

The leaves of grasses have a sheathed base. The leaf sheath usually wraps around the stem (←) above the node. Bamboo leaves (→) may have bristles (oral setae) at the sheath and the blade junction.

White millet (←), *Panicum miliaceum*

Foxtail barley (→), *Hordeum jubatum*, is a weedy species that belongs to the same genus as domestic barley.

Poaceae species illustrated on the opposite page

Oryza sativa L.—rice
Triticum aestivum L.—common bread wheat
Zea mays L.—corn, sweet corn, popcorn

The Grass Revolution

It is hard to imagine a landscape with no grass, but grasses didn't appear until late in the Cretaceous period of Earth history, which ended 65 million years ago. It wasn't until much later that grasses made up large portions of plant communities. About 20 million years ago, the grasslands biome and grass-dominated plant communities appeared. At that time, the climate became cooler and drier, which probably gave grasses a survival advantage. Today, these flowering plants make up a large portion of the terrestrial flora and are a major food source for animal life on land.

Grasses have several traits that helped them dominate plant communities. Many are drought resistant, and they also stand up well to fires and grazing. Their extensive root systems are part of this, but their special meristems also give them outstanding advantages. Meristems are the growing points of plants. Unlike animals, plants grow in only certain locations of their bodies. For most plants, the meristems that produce new shoots are at the tips of the branches. Grasses, however, have their meristems inserted at the base of the leaves, sometimes even below ground level. This means that when the leaves are eaten or cut off, the plant continues to grow.

Grazing mammals evolved in concert with grasses and developed teeth with thicker, harder enamel that could withstand the silica that grasses deposit in their leaves. For humans, grasses have a special importance. They provide us with a large portion of our daily food calories. We seldom eat the foliage, since we lack digestive systems to extract nutrients from it, but grass seeds are another matter. They provide us with carbohydrates, proteins, vitamins, and other nutrients.

Wheat, *Triticum aestivum*, grows best in cool, temperate climates. It is shown below late in flowering (left photo) and as mature, dry fruits (second from left). The "seeds" are actually the whole fruits—the pericarp adheres closely to the seed. Together, seed and tight pericarp make up the grain. Most cultivated varieties of wheat have grains that are easily threshed, removed from the remains of the bracts that surrounded their flowers.

Corn, *Zea mays*, is a monoecious grass. The tassels on the top of the stalk bear the stamens. The ear of corn (below, second from right) is the pistillate inflorescence. A kernel of corn is the ovary of a single pistillate flower. Each kernel has its own silk, which is a long style with a stigma at the end. The corn shucks are green bracts that cover this inflorescence. We eat the immature grain as sweet corn.

Rice (below, far right), *Oryza sativa*, grows in tropical to warm temperate climates. Rice produces best when it is cultivated in partly flooded fields, although it can also be grown on dry ground.

Monocots

Commelinid monocots

Poales Small

Typhaceae Jussieu   1 genus/8-13 species

Sole genus:

Typha L.—cattail

Typha latifolia L.—common cattail

Typha laxmannii Lepechin—graceful cattail, Laxman's bulrush

* illustrated

Typhaceae, The Cattail Family

Typha is the sole genus of the cattail family. These aquatic plants are monoecious. Their staminate flowers form above the pistillate flowers on the same stalk. The flowers are so small that it is hard to see any of their details. The staminate flowers fall off after they bloom, and the ovaries of the pistillate flowers develop into the familiar cattail. The common cattail, pictured below, is *Typha latifolia*.

(↓) The young flowers are green. (↓) When the staminate flowers mature, they give off large amounts of pollen. The pistillate flowers turn brown and enlarge (↓) as the seed develops while the staminate flowers wither. In the fall, the seeds are shed and blow away with the aid of attached fluffy fibers (↓).

Typha laxmannii (→), commonly called graceful cattail, is a smaller plant with narrower leaves than *T. latifolia*. It is a Eurasian native that is grown as an ornamental.

Monocots

Commelinid monocots

Zingiberales Grisebach

Cannaceae Jussieu   1 genus/19 species

Genus:

Canna L.—canna

   Canna indica L.—Indian-shot

   Canna x *generalis* L. H. Bailey—hybrid canna

* illustrated

Introducing Zingiberales, the Ginger Order

The families of the ginger order have leaves that look much alike, so much so that they can be mistaken for one another. The typical leaf has a strong midrib. The secondary veins are not parallel to the midrib, but are parallel to each other. They leave the midrib at an angle, from acute to about 90°. This pattern of veining is sometimes called pinnate-parallel. The leaves are rolled into a tube when they form; they uncoil as they mature. One sometimes sees leaves with a row of holes going across the blade, where an insect chewed through all the layers while the leaf was rolled. The flowers have inferior ovaries and no more than five functional stamens. The prayer plant and canna families have only a single functional stamen. The remaining stamens are staminoides, modified stamens that often look and function like petals. Most of the families produce aerial stems only when they bloom. The one tropical family that is not on this tour, Costaceae, grows a spiraling aerial stem with one row of leaves.

Cannaceae, The Canna Family

Cannas have showy, asymmetric flowers. They are native to Mexico and Central America, but are widely used as ornamentals. They grow from rhizomes that can be dug and stored inside during winter in temperate areas. Hybrids with large flowers (↓) and a broad range of colors are commonly grown.

It isn't easy to figure out the structure of a canna flower, but the parts show up more clearly on the species, *Canna indica* (→). There are typically three or four modified stamens that look like petals, and which are called staminodes. The one functional stamen has half an anther and is also petal-like. On the pistil, the style is flattened. The three true petals are slender and pointed. Their bases are fused, along with the stamens and the style, and form a tube at the base of the flower. Their fruit is a three-sided capsule that is covered with pointy bumps (↓).

(←) The calyx remains on the end of the developing fruit.

Monocots

Commelinid monocots

Zingiberales

Heliconiaceae Nakai   1 genus/100-200 species

Sole genus:

Heliconia L.

Heliconia orthotricha L. Andersson—heliconia

Heliconia psittacorum L. f.—parrot heliconia

Heliconia rostrata Ruiz & Pavón—lobster claw, fishpole heliconia

* illustrated

Heliconias have groups of flowers inside large, brightly colored bracts. The flowers themselves are smaller and much less showy than the bracts. Heliconias are tropical plants from the Americas. Their leaves and bracts are two-ranked, which means they are in two rows on either side of the stem.

Heliconia orthotricha (↑) and *H. rostrata* (↑) are grown as ornamentals in many tropical or subtropical areas.

The bracts of this parrot heliconia (↑), *H. psittacorum*, are red. The flowers have tubular yellow tepals with dark green tips.

The developing fruits of this parrot heliconia (↓) have three sections because they develop from a three-carpellate ovary. Only one seed forms in each locule. A three-chambered ovary that has one ovule in each chamber is characteristic of heliconias.

Monocots

Commelinids

Zingiberales

71

Monocots

Commelinid monocots

Zingiberales

Marantaceae R. Brown   31 genera/550 species

Genera include:

Calathea G. Meyer

   Calathea makoyana E. Morrin—peacock plant

Ctenanthe Eichler—ctenanthe

Maranta L.—prayer plant, arrowroot

   Maranta leuconeura var. *erythroneura* G. S. Bunting

   Maranta leuconeura var. *kerchoviana* (E. Morrin) Petersen

Stromanthe Sonder—stromanthe

Thalia L.—water canna

* illustrated

Marantaceae, The Prayer Plant Family

Maranta, overview of foliage (↑) and close-up view of the small flowers (↑)

The leaves of this family fold upward at night, which gives them the common name of prayer plant. They are native to subtropical and tropical areas worldwide. In temperate areas, they are houseplants, grown for their beautifully marked leaves. The pinnate-parallel venation of the order shows clearly in these leaves. The flowers are asymmetric and emerge from bracts.

Calathea Corona (←) and *Calathea mako-yana*, known as peacock plant (↑)

Maranta leuconeura var. *kerchoviana* (↓, ↘) is called rabbit tracks. Its flower is much smaller than the leaves. Red herringbone plant (←), *Maranta leuconeura* var. *erythroneura*, is another variety of this species.

73

Monocots

Commelinid monocots

Zingiberales

Musaceae Jussieu   2 genera/35 species

Genera are:

Ensete Horaninow—Abyssinian banana

*Musa L.—banana, plantain (tropical banana-like fruit)

    *Musa ornata Roxburgh—ornamental banana

    *Musa velutina H. Wendland & Drude—velvet banana, self-peeling banana

* illustrated

Ornamental banana
(→), *Musa ornata*

The banana, genus *Musa*, has large red or purple bracts that surround its small, tubular flowers. The flowers have yellow or cream-colored tepals. The pistillate flowers bloom first in the inflorescence. (←) Pistillate flowers have inferior ovaries, which are green in this example. Later flowers (↑) are staminate. This produces a bunch of bananas and a long section of bare stem where the staminate flowers once bloomed. The bananas from the grocery store do not form seeds. Consequently, they are propagated vegetatively from the plant's corms. The banana's leaves spiral around the stem.

The velvet banana (→), *Musa velutina*, is grown as an ornamental. At the top of the shoot, the staminate flowers are blooming. Their yellow tepals can be seen above the curled red bract. At the base of the inflorescence, the pistillate flowers have already bloomed and their ovaries have begun to develop into the fruits. These small, red fruits are not edible. They have a thick peel (↓) and a fibrous center that is full of very hard, black seeds.

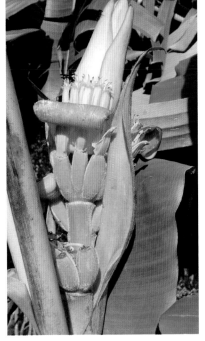

Monocots

Commelinid monocots

Zingiberales

Strelitziaceae Hutchinson   3 genera/7 species

Genera include:

Ravenala Adanson—traveler's palm

Strelitzia Aiton

　　Strelitzia nicolai Regel & Koernicke—giant bird-of-paradise,
　　　white bird-of-paradise

　　Strelitzia reginae Aiton—bird-of-paradise

* illustrated

Strelitziaceae, The Bird-of-paradise Family

The flowers of this family are so modified that finding their parts is not easy. A series of flowers grow along the top of a large horizontal bract. The sepals of bird-of-paradise, *Strelitzia reginae* (↑), are bright orange. The petals are blue (↑). One petal is separate; it is the small blue structure that stands near the base of the sepals (white arrow). The rest of the petals are fused together and wrap around the long, pointed style and the five stamens. The stamens are fused inside a groove in the long, blue petals. The ovary has three chambers, each of which holds numerous ovules.

The giant bird-of-paradise, *Strelitzia nicolai*, has sheathing leaf bases. The leaves (↑) form two rows (ranks) on either side of the stem. The flowers (↑) have a large, dark purplish blue bract at their base. Its sepals are white to pale pink. This showy commelinid monocot is also known as the white bird-of-paradise.

The other major genus of this family is *Ravenala*, the traveler's palm of Madagascar, which is grown for its striking form, a large fan of leaves that looks much like giant bird-of-paradise.

Monocots

Commelinid monocots

Zingiberales

Zingiberaceae Martynov   46-52 genera/1075-1300 species

Genera include:

Alpinia Roxburgh—shell ginger, galangal

     Alpinia zerumbet (Persoon) B. L. Burtt & R. M. Smith—shell ginger

Curcuma L.—Siam tulip, East Indian arrowroot, mango ginger

     Curcuma longa L.—tumeric

Elettaria Maton

     Elettaria cardamomum (L.) Maton—cardamom

Etlingera Giseke—torch ginger, waxflower

     Etlingera elatior (Jack) R. M. Smith—torch ginger

Hedychium Koenig—ginger lily

Kaempferia L.—peacock ginger

Zingiber P. Miller—ginger

     Zingiber officinale Roscoe—culinary ginger

* illustrated

Zingiberaceae, The Ginger Family

This family has irregular flowers with a complicated structure. All but one of the stamens have become petal-like staminodes, such as the ruffled yellow and red part in the *Alpinia* flower (→). The filament of the single functional stamen (arrow) is broad and folds around the style. The three petals of this shell ginger, the bract at the flower's base, and the calyx are all white.

The stripes of variegated shell ginger, *Alpinia zerumbet* 'Variegata', (→, ↘) show the pinnate-parallel venation pattern of the leaves.

Curcuma longa (↑) has pink bracts that hold small yellow and white flowers.

The ginger family is the source of several important seasonings. The fruits of cardamom, *Elettaria cardamomum*, are picked slightly green and then dried. The seeds are used to flavor baked goods, Indian foods, coffee, and tea. The pungent rhizomes of ginger, *Zingiber officinale*, are used as a spice and a medicine. It has been cultivated in Asia from prehistoric times. Turmeric rhizomes, from *Curcuma longa*, are used as a flavoring, a preservative, and a dye.

Torch ginger, *Etlingera elatior* (↑), is a striking ornamental in tropical gardens. Its flowers are the small red structures with yellow edging.

79

Tree Diagram of Flowering Plant Orders: Eudicots

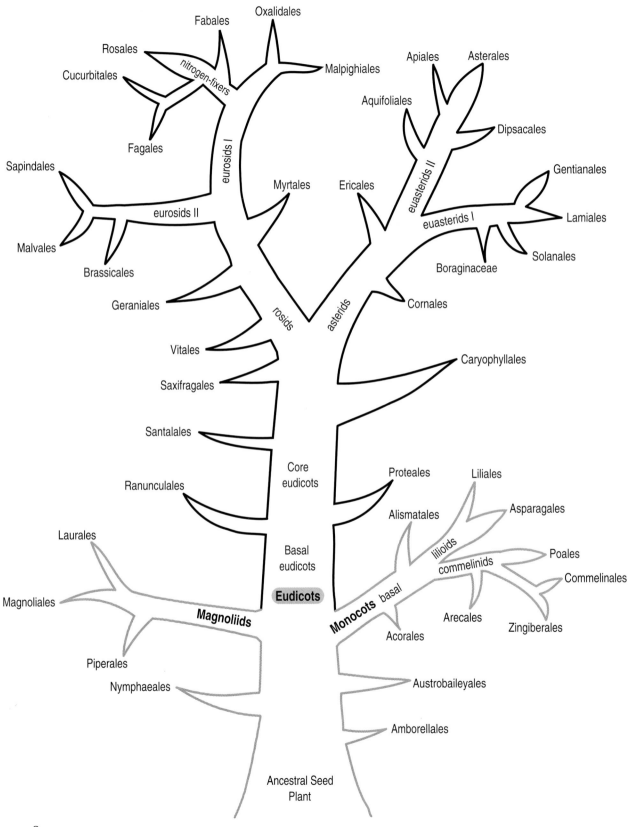

Introduction to the Eudicots

Traditionally, the dicots were one of two main groups of flowering plants. They were named for their two cotyledons or seed leaves, which can be seen in the photo above of a sunflower seedling. The dicots are now known to be a mixture of plant groups descended from different ancestors. The eudicots are what remain after the magnoliids and other basal angiosperms are removed from the traditional dicots. The name "eudicot" literally means "true dicots."

One characteristic common to eudicots is pollen with three openings. An alternate name for the group is the tricolpates, from the term for pollen grains with three openings. Eudicot flowers typically have parts such as petals and stamens in multiples of four or five. This rule doesn't apply to the pistils, which may be a single carpel, two-carpellate, or three-carpellate or more.

The leaves of eudicots typically have netted veins; they generally do not have parallel veins. After that, there are many variations, such as simple or compound leaves, with or without stipules, and with a wide diversity of lobes and margins.

This is by far the largest branch of the flowering plants, as it holds about three quarters of the species. Eudicots are an ancient group, whose history can be tracked via its characteristic pollen. The first eudicot fossil pollen appears in 125 million-year-old rocks. In comparison, the oldest fossils we have that are clearly flowering plants are just a little older, 130 million years old.

Eudicot plant forms and adaptations cover the full diversity of flowering plants, as they include trees, shrubs, vines, herbaceous plants, aquatics, succulents, carnivorous plants, and parasitic plants.

The first branch of the eudicots is the order Ranunculales, which includes the buttercup family (Ranunculaceae), the poppy family (Papaveraceae), and the barberry family (Berberidaceae). The next branch is the Proteales, which includes the lotus family (Nelumbonaceae) and the sycamore family (Platanaceae). These orders are considered basal eudicots.

The rest of the group are called the core eudicots. They split into five main groups—the Caryophyllales, the Santalales, the Saxifragales, and the two biggest branches, the rosids and the asterids. There is more information about these five groups in their sections of this tour.

Eudicots

Basal eudicots

Ranunculales Dumortier

Berberidaceae Jussieu   14 genera/700 species

Genera include:

Achlys DC.—vanilla leaf, deerfoot

Berberis L.—barberry

    **Berberis thunbergii* DC.—Japanese barberry

Caulophyllum Michaux—blue cohosh

Diphylleia Michaux—umbrella leaf

Epimedium L.—epimedium, bishop's hat, barrenwort

    *Epimedium rubrum C. Morren—red epimedium

Jeffersonia Barton—twinleaf

Mahonia Nuttall—mahonia, holly grape

    **Mahonia aquifolium* (Pursh) Nuttall—Oregon grape

Nandina Thunberg

    **Nandina domestica* Thunberg—nandina, heavenly bamboo, sacred bamboo

Podophyllum L.—may apple, American mandrake

Vancouveria Morren & Decaisne—inside-out flower

* illustrated

Berberidaceae, The Barberry Family

Oregon grape (↑), *Mahonia aquifolium*, flowers and fruits (↑)

The barberry family blooms in spring. Its shrubs have yellow wood, colored by a substance called berberine. The flowers have two whorls of perianth, with four to six outer petal-like sepals surrounding six shorter petals. There is a single carpel in the simple pistil. The fruits are usually berries. The leaves of many species are compound and many have spines on the leaf margins. Some species also have spines on their stems.

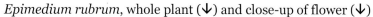

Epimedium rubrum, whole plant (↓) and close-up of flower (↓)

The Japanese barberry (↓), *Berberis thunbergii*, has many horticultural forms. The variety called Atropurpurea (1) has dark red foliage. The fruits (2) remain on the branches in winter.

Nandina (↓), *Nandina domestica*, also called heavenly or sacred bamboo, is frequently used in landscaping. It is unrelated to true bamboo, which is in the grass family. Its leaves are bipinnately compound, with triple and single leaflets. The flower has six stamens and a carpel (↓) with a knob-like stigma. Nandina originated in eastern Asia.

Eudicots

Basal eudicots

Ranunculales Dumortier

Papaveraceae Jussieu   41 genera/760 species

Genera of Papaveroideae, the poppy subfamily, include:

Argemone L.—prickly poppy

    *Argemone polyanthemos (Fedde) G. B. Ownbey—crested prickly poppy

Chelidonium L.—celandine, swallowwort

Dendromecon Bentham—tree poppy

Eschscholzia Chamisso

    *Eschscholzia californica Chamisso—California poppy

Macleaya R. Brown—plume poppy

Meconopsis Viguier—blue poppy, Himalayan poppy, Welsh poppy

Papaver L.—poppy, alpine poppy, opium poppy, Shirley poppy

    *Papaver bracteatum Lindley—oriental poppy

    *Papaver nudicaule L.—Iceland poppy

Romneya Harvey—California tree poppy, matilija poppy

Sanguinaria L.—bloodroot

Stylophorum Nuttall—wood poppy, celandine poppy

Genera of Fumarioideae, the fumitory subfamily, include:

Corydalis DC.—corydalis, fumewort

    *Corydalis aurea Willdenow—golden smoke, scrambled eggs, golden corydalis

Dicentra Bernhardi—bleeding heart

    *Dicentra spectabilis (L.) Lemaire—common bleeding heart

Fumaria L.—fumatory

* illustrated

(↑) Opening bud, flower, and fruit of Iceland poppy, *Papaver nudicaule*

Poppies' two green fuzzy sepals fall off when the flower blooms. Their four to six wrinkled petals unfold, revealing the many stamens and the compound pistil. The stigma sits on top the ovary; there is little or no style. The fruit is a capsule, often with many holes near the top, like a salt shaker on a stalk. Poppies often have milky sap that is white or colored.

California poppy (↓),
Eschscholzia californica

Oriental poppy (↓),
Papaver bracteatum

Prickly poppy (↓),
Argemone polyanthemos

The fumitory subfamily has distinctive flowers with bilateral symmetry. The fanciful names include bleeding hearts, squirrel corn, and Dutchman's breeches.

Common bleeding heart (↓), *Dicentra spectabilis*

Golden corydalis (↓), *Corydalis aurea*

85

Eudicots

Basal eudicots

Ranunculales Dumortier

Ranunculaceae Jussieu   62 genera/2525 species

Genera include:

Aconitum L.—monkshood

    **Aconitum carmichaelii* Debeaux—Chinese monkshood

Actaea L.—baneberry, black cohosh (formerly *Cimicifuga*)

*Anemone L.—windflower, anemone, hepatica (formerly *Hepatica* Miller), pasque flower (formerly *Pulsatilla* Miller)

    *Anemone x *lesseri* Wehrhahn—windflower

Aquilegia L.—columbine

    **Aquilegia caerulea* E. James—Colorado columbine

Caltha L.—marsh marigold

*Clematis L.—clematis, virginsbower

    **Clematis tangutica* (Maximowicz) Korshinsky—golden clematis

Coptis Salisbury—goldthread

Coriflora W. A. Weber—leather flower, sugarbowls

Delphinium L.—delphinium, larkspur

Eranthis Salisbury—winter aconite

Helleborus L.—Lenten rose, Christmas rose, hellebore

    **Helleborus orientalis* Lamark—Lenten rose

Hydrastis L.—goldenseal

Nigella L.—nigella, love-in-a-mist

Ranunculus L.—buttercup, spearwort, blisterwort, crowfoot

Thalictrum L.—meadow rue, rue anemone, king of the meadow

Trollius L.—globe flower

* illustrated

Ranunculaceae, The Buttercup Family

Colorado columbine (←), *Aquilegia caerulea*, flowers and developing follicles (left photo). The flowers have blue sepals. The white petals have a long nectar spur (white arrow). Each separate carpel becomes a follicle.

Chinese monkshood (→), *Aconitum carmichaelii*

The flowers of Ranunculaceae have many stamens. Most have many separate carpels. All shown here except *Aquilegia* lack petals and have only colored sepals. Family members have many kinds of fruits. Columbines, delphiniums, and monkshood have clusters of follicles. Clematis has bunches of achenes with long, feathery styles. Some anemones have a rounded aggregate of carpels (↓) that breaks into many fuzzy achenes. Baneberries have racemes of berry-like fruits. The leaves are often compound or lobed. Monkshood and baneberry are highly poisonous.

This *Clematis* hybrid (↓) has purple anthers surrounding the white stigmas. The carpels each produce an achene with a long, feathery style.

The flower and developing fruits of *Anemone* x *lesseri* (↑) show a typical aggregate of carpels that are closely grouped, but remain distinct.

The developing fruits of golden clematis, *C. tangutica* (→)

Lenten rose (←), *Helleborus orientalis*, has red styles and yellow stamens.

87

Eudicots

Basal eudicots

Proteales Dumortier

Nelumbonaceae Berchtold & J. Presl   1 genus/2 species

Genus:

Nelumbo Adanson

Nelumbo lutea Persoon—American lotus, water chinaquin

*Nelumbo nucifera Gaertner—sacred lotus, Chinese lotus

* illustrated

Relatives Do Not Always Look Alike and Look-alikes May Not Be Related

If you look at older classifications and field guides, you will likely see the lotus, *Nelumbo*, placed in the water lily family, Nymphaeaceae. Both are aquatics with large showy flowers and rounded leaves. The flowers both have a spiral arrangement of parts. They are not related, however. Their similarities are the result of convergent evolution. They may look the same because of the environment in which they live and the similar adaptations it selects, but they do not share a common ancestor.

The lotus and the water lily are a good example of the problems of classifying on structures alone. More data, especially the crucial DNA data, show that water lilies are one of the early branches of the flowering plants. Water lilies' only close relatives are another family of aquatic plants, the fanworts, Cabombaceae.

The lotus is much further up our tree diagram of flowering plants, in the basal eudicots. Its closest relatives are the sycamore trees of the family Platanaceae and Proteaceae, the protea family, which occurs mainly in the Southern Hemisphere.

A lotus flower may not look much like a sycamore flower, but they share a common ancestor. Since their relationship was discovered, it has been possible to find traits they have in common. These include details of seed development, the internal structure of their vessels, and the form of the waxes on their leaves. Plants have a myriad of structures. Without DNA evidence, it is very difficult to know which structures indicate close relationships.

Nelumbo is the sole genus in this family of aquatic plants. It holds only two species, *N. nucifera*, the sacred lotus from Asia (←), and *N. lutea*, the American lotus, which has yellow flowers.

The leaves of the sacred lotus have their petioles attached in the middle of the disk-shaped leaf blade, an arrangement that is called a peltate leaf. The leaves are held above the water's surface.

The flower of the sacred lotus has two green sepals and numerous petals, some of which are greenish white (←, →). Its enlarged receptacle has an inverted cone shape. The many stamens and petals attach at the receptacle base (↙). The many ovaries are embedded into the surface of the receptacle (↓). There are no styles; the stigma covers the top of each ovary. The fruits (↘) develop in pockets on the receptacle surface.

Eudicots

Basal eudicots

Proteales Dumortier

Platanaceae T. Lestibudois   1 genus/10 species

Genus:

Platanus L.—sycamore

  **Platanus occidentalis* L.—American sycamore

  **Platanus racemosa* Nuttall—California sycamore

* illustrated

Platanaceae, The Sycamore Family

The sycamore or plane tree can be recognized by the white or tan patches on its trunk, which are revealed by flaking bark. The flowers are tiny and borne in little balls. Pistillate and staminate flowers are in separate inflorescences, but occur on the same tree. The stipules at the base of young leaves resemble little leaves that grow completely around the stems. *Platanus* is the sole genus of this family.

The spherical inflorescence becomes a ball of achenes (↑), along with bristles that help blow the fruits to new locations. The leaf petioles completely cap the axillary buds.

American sycamore (↑),
Platanus occidentalis

Petiole of shed leaf next to a bud (↑). The petiole base (white arrow) is hollow because it completely surrounded the bud. The leaf scar that is left (black arrow) encircles the dormant bud.

(↑) Young pistillate inflorescence and (↑) developing fruit head of the California sycamore, *Platanus racemosa*

Eudicots

Core eudicots

Caryophyllales Perleb

Aizoaceae Martynov   123 genera/2020 species

Genera include:

Aptenia N. E. Brown

    *Aptenia cordifolia (L. f.) N. E. Brown—heartleaf ice plant

Carpobrotus N. E. Brown—sea fig, Hottentot fig

Celphalophyllum (A. H. Haworth) N. E. Brown—red spike ice plant

Delosperma N. E. Brown—ice plant

    *Delosperma cooperi L. Bolus—hardy ice plant

    *Delosperma nubigenum (Schlechter) L. Bolus—hardy yellow ice plant

Dorotheanthus Schwantes—Livingstone daisy

Fenestraria N. E. Brown

    *Fenestraria rhopalophylla N. E. Brown—baby toes

Glottiphyllum Haworth—tongue leaf plant

Lampranthus N. E. Brown—ice plant

*Lithops N. E. Brown—living stones

Malephora N. E. Brown—ice plant

Mesembryanthemum L.—ice plant

Oscularia Schwantes—oscularia

Tetragonia L.—New Zealand spinach

Titanopsis Schwantes—concrete leaf, little tortoise foot

* illustrated

Introducing Caryophyllales

The order Caryophyllales holds approximately 6% of the eudicots. Although this branch is far smaller than the rosids or asterids, it holds a wide variety of plant forms and adaptations. Included are succulents and the cacti, both champions at life in arid climates. Several groups, including ice plants, are able to grow in salty soils, another environment that challenges the plant's ability to retain water.

Several families of insectivorous plants are part of Caryophyllales. These include the Venus fly trap family, Droseraceae, and the Asian pitcher plant family, Nepenthaceae.

Many families in this order have a unique type of pink pigments called betalains. These pigments provide the intense red of beets and the vivid pink of cactus flowers. Some families, including Caryophyllaceae, have anthocyanins, the same pink pigments as most of the other flowering plants.

One thing all Caryophyllales share is simple, entire leaves. The flowers are much more variable than the leaves. The flowers are often complex and deceptive in structure. The calyx may be colored and take over the function of petals. Even when the flowers have petals and sepals, it seems these structures do not develop in the same way as they do in other flowering plants.

Aizoaceae, The Ice Plant Family

The ice plant family has thickened, succulent leaves. This is an adaptation for growing in dry habitats or in salty soils, such as those near the ocean. The leaves are opposite, and their bases wrap around the stem. The flowers of this family look similar to daisies, but the structure is completely different. There are many stamens and petals and one pistil that has an inferior ovary with four to seven locules. There is a stigma for each locule. The fruits of some genera open only when they get wet.

Delosperma nubigenum (←) and *D. cooperi* Kelaides (↓) are among the hardiest of ice plants, meaning that they can survive lower temperatures than most. Many ice plants cannot survive freezing!

The flower of *Delosperma cooperi* has a ring of sterile, stamen-like structures surrounding its functional stamens. The five stigmas (↑) and (↑) the five-part symmetry of the fruit show that this species has a five-carpellate pistil.

Fenestraria rhopalophylla (→) is adapted to life in extremely dry conditions. The tip of each fleshy leaf is transparent, providing a little window into the leaf's interior.

Heartleaf ice plant (←), *Aptenia cordifolia,* has succulent leaves, but thinner ones than most of this family have.

Lithops species (→) are called living stones for good reason. Growers of succulents cultivate many varieties of these plants. Only the tops of the two succulent leaves of *Lithops* protrude above ground.

93

Eudicots

Core eudicots

Caryophyllales Perleb

Amaranthaceae Jussieu   174 genera/2050 species

Genera include:

Alternanthera Forsskaol—joyweed, calico plant, khaki weed

Amaranthus L.—amaranth, pigweed, Joseph's coat, love-lies-bleeding

    **Amaranthus hypochondriacus* L.—prince's feather

    **Amaranthus palmeri* S. Watson—careless weed, redroot pigweed

Atriplex L.—saltbush

Beta L.—beets, chard

Celosia L.—cockscomb

Chenopodium L.—goosefoot, lambsquarters, strawberry blite, quinoa, epazote

    **Chenopodium capitatum* (L.) Ambrosi—strawberry blite, blite goosefoot

Froelichia Moench—cottonweed

Gomphrena L.—globe amaranth

    **Gomphrena globosa* L.—common globe amaranth

Iresine P. Browne—bloodleaf

Kochia Roth—kochia, molly, summercypress

    *Kochia scoparia (L.) Schrader—kochia, Mexican fireweed

Salicornia L.—glasswort, pickleweed, saltwort, swampfire, samphire

Salsola L.

    *Salsola kali L.—Russian thistle, tumbleweed

Spinacia L.

    **Spinacia oleracea* L.—spinach

NOTE: This family includes the former Chenopodiaceae, the goosefoot family, which is no longer recognized.

* illustrated

Amaranthaceae, The Amaranth and Goosefoot Family

The flowers of this family are very small and hard to see, although many members have large inflorescences, made showy by their colored bracts. Among the many weedy species are:

Careless weed or redroot pigweed (←), *Amaranthus palmeri*

Kochia (→), *Kochia scoparia*

Strawberry blite (↙), *Chenopodium capitatum*

Tumbleweed or Russian thistle (↓,↘), *Salsola kali.*

Tumbleweed or Russian thistle blooms in the fall. (↑) Its flowers have pink or white sepals. The whole plant breaks off in late autumn and rolls away in the wind, scattering its abundant seeds.

Red seed coverings (↑) give strawberry blite its name.

Spinach (↓), *Spinacia oleracea,* is dioecious, which means it has staminate and pistillate flowers on separate plants.
(↓) Staminate plant and pistillate plant (↓)

Ornamentals include *Gomphrena globosa,* globe amaranth (↓), which has pink bracts and tiny yellow flowers. Prince's feather (↓), *Amaranthus hypochondriacus,* has a tall inflorescence of tiny colored bracts and numerous, minute flowers.

Eudicots

Core eudicots

Caryophyllales Perleb

Cactaceae Jussieu　　　　　　100 genera/1500 species

Genera include:

Carnegiea Britton & Rose—saguaro cactus

Cereus P. Miller—queen-of-the-night, giant club cactus, hedge cactus

Echinocactus Link & Otto—barrel cactus, devil's-head cactus

　　Echinocactus texensis Hopffer—devil's-head cactus, horse-crippler cactus

Echinocereus Engelmann—hedgehog cactus

　　Echinocereus fendleri (Engelmann) F. Seitz—Fendler's hedgehog cactus

Ferrocactus Britton & Rose—barrel cactus

Hylocereus (A. Berger) Britton & Rose—night-blooming cereus

Lophophora J. M. Coulter—peyote

Mammillaria Haworth—fishhook cactus, nipple cactus, pincushion cactus

Opuntia P. Miller—prickly pear, cholla

　　Opuntia phaeacantha Engelmann—tulip prickly pear

　　Opuntia polyacantha Haworth—plains prickly pear, starvation cactus

Pediocactus Britton & Rose—ball cactus

Schlumbergera Lemaire—Christmas cactus, orchid cactus, crab cactus

　　Schlumbergera truncata (Haworth) Moran—Thanksgiving cactus, crab cactus

* illustrated

Is It a Cactus or Not?

Many plants that are adapted to harsh desert climates share a number of features. They often have very small leaves or no leaves. Many have transferred the function of photosynthesis to their green stems, which also serve as water storage structures. These plants are not all cacti, however.

The cactus family, except for one member, is native to the Western Hemisphere. Cacti have spines, structures that are derived from leaves. Their spines grow in clusters on structures called areoles. Other desert plants may have prickly structures that help defend them from herbivores, but either those structures are not true spines or they are not arranged on areoles. The flowers of cacti are also distinctive and are key to identifying a plant as a cactus.

Cactus-like desert plants from the Eastern Hemisphere belong to a number of families, including Euphorbiaceae, Apocynaceae, and even Asteraceae. Euphorbia family members often have a thick, milky sap. The common name of pencil cactus is given to some euphorbias, but their milky sap and lack of spines show that they are not true cacti. The flower structure is the best way to identify the family of a cactus-like succulent.

The cacti and very similar non-cacti show that common structures can be the result of adaptations for the same environmental stresses. This process is called convergent evolution. Structure alone is not reliable information for classifying plants, but it can certainly be part of the data. The Angiosperm Phylogeny Group classification uses many types of information, including DNA comparison and plant developmental information, to place flowering plants in related groups.

The plains prickly pear (↑), *Opuntia polyacantha*, can bear either pink or yellow flowers. (↑) The pinkflower hedgehog cactus, *Echinocereus fendleri*, grows in Arizona, New Mexico, Texas, and Colorado.

Most cacti are easily recognized by their thickened, fleshy stems that are covered with areoles, which are seen as patches of spines and prickly hairs. Many cacti have ribbed stems that can expand to hold water after the infrequent rains of their desert habitats. Some cacti have lost their spines and live as epiphytes in tropical forests. The spines of cacti are modified leaves, the only remnants of leaves they have. Only a few, very primitive cacti have recognizable leaves. The stems have taken over the job of photosynthesis.

Cactus flowers (↑) have many stamens and many tepals. The tepals gradually change (↑) from sepal-like to petal-like from the outside to the inside of the flower. There is a single style with a lobed or branched stigma, which is green in the prickly pear photo above. The inferior ovary is embedded in the floral shoot. Together they develop into the fruit. (↙) Cactus fruits are typically fleshy, colorful berries.

The devil's head or horse-crippler cactus (→), *Echinocactus texensis*, is native to the short-grass prairies of Texas and New Mexico. Its stout spines can puncture a car tire.

A prickly pear (↑), *Opuntia phaeacantha*, with red fruits

Schlumbergera truncata (→) is commonly called Thanksgiving cactus or crab cactus. Its flowers bud from the flattened stem segments. There are no leaves or spines. This group of cacti is one of the few that have bilaterally symmetrical flowers instead of radially symmetrical ones.

Eudicots

Core eudicots

Caryophyllales Perleb

Caryophyllaceae Jussieu   86 genera/2200 species

Genera include:

Agrostemma L.—corn cockle

Arenaria L.—sandwort

Cerastium L.—chickweed, mouse-ear

    **Cerastium tomentosum* L.—snow-in-summer

Dianthus L.—pink, sweet william, carnation

    **Dianthus chinensis* L.—china pink

Gypsophila L.—baby's breath, gypsophila

Lychnis L.—campion, catchfly

    **Lychnis chalcedonica* L.—Maltese cross

Minuartia L.—sandwort

Sagina L.—pearlwort, Irish moss, Scotch moss

Saponaria L.—soapwort, bouncing bet

    **Saponaria ocymoides* L.—rock soapwort

Silene L.—campion, wild pink, royal catchfly, fire pink

    *Silene latifolia Poiret ssp. *alba* (P. Miller) Greuter & Burdet—white campion, white cockle

Stellaria L.—chickweed, starwort

Vaccaria Wolf—cow cockle

* illustrated

98

China pink (←), *Dianthus chinensis*
Maltese cross (↑), *Lychnis chalcedonica*
Rock soapwort (↓), *Saponaria ocymoides*
Snow-in-summer (←), *Cerastium tomentosum*

Pink family members usually have enlarged nodes with opposite leaves. The petals of this family typically have notches or "pinking." The petals are clawed, which means that their bases narrow markedly. The pistils range from two to five carpels with the ovary fused, but the stigmas and styles free. The fruit is a capsule.

(↑) The five petals of the white campion flower, *Silene latifolia* ssp. *alba*, are so deeply notched that they look like twice that number. A tubular calyx encloses the base of the petals. This species is dioecious. In the pistillate flower, the young green ovary (↑), as revealed by peeling back the perianth, is topped by its five separate styles and stigmas. As the fruit matures, the green ovary (↑) enlarges inside the striped, persistent calyx, which was peeled back for the photo. The mature capsule has a ring of teeth (↑) that fold back from the top, exposing the loose seeds in the cup-shaped, dry ovary.

Eudicots

Core eudicots

Caryophyllales Perleb

Droseraceae Salisbury   3 genera/115 species

Genera are:

Aldrovanda L.—waterwheel plant

Dionaea Ellis

    **Dionaea muscipula* Ellis—Venus flytrap

Drosera L.—sundew

    *Drosera binata Labillardière—forked leaf sundew

    *Drosera capensis L.—Cape sundew

* illustrated

Droseraceae, The Venus Flytrap Family

Venus flytrap (↑), *Dionaea muscipula*

Cape sundew (→), *Drosera capensis*

The Droseraceae are one of several families of insectivorous plants in the order Caryophyllales. These plants have remarkable adaptations for living in boggy, nitrogen-poor soils. They trap insects, which provide a supplemental source of nitrogen and other nutrients.

The Venus flytrap has leaves modified as closable traps. When an insect touches two or more of the tiny hairs inside the trap, it closes and kills the insect. Cell growth is required to open and close the trap. Each trap can close about five times before it no longer works.

The sundew has sticky hairs on its leaves. The long hairs capture insects and then bend inward, bringing the insect close to the leaf, where nutrients are absorbed as the insect's body breaks down.

There are five petals, five stamens, and a three-carpellate pistil in flowers (↓) of *Drosera binata*. The leaves of the cultivar 'Multifida' branch twice (↓).

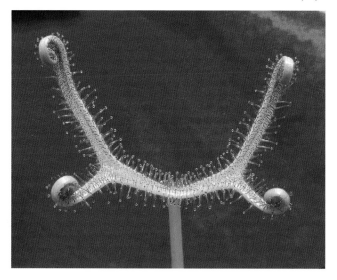

Eudicots

Core eudicots

Caryophyllales Perleb

Nyctaginaceae Jussieu   30 genera/395 species

Genera include:

Abronia Jussieu—sand verbena

   *Abronia fragrans Nuttall ex Hooker—snowball sand verbena

Acleisanthes A. Gray—angel trumpets

Allionia L.—trailing four o'clock, windmills

Boerhavia L.—spiderling

*Bougainvillea Commerson ex Jussieu—bougainvillea

Mirabilis L.—four o'clock, umbrellawort, wishbone-bush

   *Mirabilis jalapa L.—garden four o'clock, marvel of Peru

   *Mirabilis multiflora (Torrey) A. Gray—wild four o'clock, Colorado four o'clock

   *Mirabilis rotundiflora (Greene) Standley—roundleaf four o'clock

Nyctaginia Choisy—devil's bouquet

Oxybaphus L'Héritier ex Willdenow—umbrellawort

Selinocarpus A. Gray—moonpod

Tripterocalyx Hooker ex Standley—sandpuffs

* illustrated

Nyctaginaceae, The Four O'clock Family

The genus *Mirabilis* includes a common garden flower, the four o'clock or marvel-of-Peru, *M. jalapa* (↑), which is a native of tropical South America. Its flowers open in the late afternoon and remain open at night. This species has become naturalized in parts of the United States. It is difficult to eradicate because it produces abundant seeds.

This *Bougainvillea* hybrid (↑) has small white flowers. The larger pink structures are floral bracts. Hybrids are available with a wide range of bract color, including shades of pink, magenta, orange, and red, as well as white.

The flowers of Nyctaginaceae are deceptive. What appears to be a fused corolla is actually colored, fused sepals. There are no petals. Usually there are bracts at the base of the flower. The bracts may be green or brightly colored. The bracts of *Bougainvillea* (↑) help attract pollinators and later assist in seed dispersal. The four o'clock family is predominately tropical and subtropical, with a few temperate members.

Native species of *Mirabilis* include the wild four o'clock (↑), *M. multiflora,* which grows throughout the southwestern United States, and round-leaf four o'clock, *M. rotundifolia* (↓), which occurs only in Colorado.

Abronia fragrans (↓) is commonly called the sand verbena, but it is a member of the four o'clock family, not the verbena family.

Eudicots

Core eudicots

Caryophyllales Perleb

Plumbaginaceae Burnett   27 genera/836 species

Genera include:

Acantholimon Boissier—prickly thrift

Armeria Willdenow—thrift

　　Armeria maritima (Miller) Willdenow—sea pink, common thrift

Ceratostigma Bunge—plumbago

　　Ceratostigma plumbaginoides Bunge—dwarf or Chinese plumbago

Limonium Miller—statice

　　Limonium perezii (Stapf) F. T. Hubbard—Perez's sea lavender

　　Limonium sinuatum (L.) Miller—annual statice, sea lavender

Plumbago L.—Cape plumbago, leadwort

* illustrated

Plumbaginaceae, The Leadwort Family

Sea pink or common thrift (←), *Armeria maritima*, whole plant and close-up view of an inflorescence

Dwarf plumbago (←), *Ceratostigma plumbaginoides*

The flowers of Plumbaginaceae have five petals, five stamens, and a five-carpellate pistil. Many family members have an unusual calyx that is thin, papery, and brightly colored.

Limonium perezii is one of several salt-tolerant species that are commonly called sea lavender. Shown here is a close-up view of the flowers with their purple calyxes (↓) and white corollas. A view of the whole plant is below.

This statice (↓), *Limonium sinuatum*, has white flowers. The papery calyxes are purple in this plant, but can be pink or yellow in others. The inflorescence is often dried and used in flower arrangements.

The flower spikes of prickly thrift (↓), *Acantholimon* sp., grow from a cushion of needle-like leaves. The fruit develops within the white papery calyx, which aids in seed dispersal.

Eudicots

Core eudicots

Caryophyllales Perleb

Polygonaceae Jussieu — 43 genera/1100 species

Genera include:

Antigonon Endlicher—coral vine

Bistorta (L.) Adanson—bistort

Coccoloba P. Browne—sea grape

Eriogonum Michaux—sulfur flower, wild buckwheat, eriogonum

> *Eriogonum umbellatum* Torrey—sulfur flower

Fagopyrum Miller—buckwheat

Fallopia Adanson—silver lace vine, Japanese knotweed, black bindweed

Persicaria (L.) Miller—snakeweed, knotweed

> *Persicaria amplexicaulis* (D. Don) Ronse Decraene—mountain fleece, fleece flower

> *Persicaria microcephala* (D. Don) H. Gross—fleece flower, knotweed

Polygonum L.—knotweed, smartweed, bistort

> *Polygonum cuspidatum* Siebold & Zuccarini—Japanese knotweed

Rheum L.—rhubarb

> *Rheum* x *hybridum* Murray—garden rhubarb

Rumex L.—sorrel, dock

NOTE: *Bistorta*, *Fallopia*, and *Persicaria* may be included in *Polygonum*.

* illustrated

Polygonaceae, The Buckwheat Family

(←) *Persicaria amplexicaulis,* and *Persicaria microcephala* (→) are grown as ornamentals.

These plants are related to common knotweeds. Some botanists place them in genus *Polygonum*; others prefer the genus *Persicaria*.

Japanese knotweed (↘), *Polygonum cuspidatum*, is an invasive pest that was first imported into the United States as an ornamental.

Most members of Polygonaceae have a distinctive, thin sheath that forms a little tube around the stem above each node, although sulfur flowers, genus *Eriogonum*, lack this sheath. The flowers are small and have a perianth of five to six tepals. In some species, the perianth persists after bloom and may develop into wings or other structures around the fruit.

Sulfur flower or sulfur buckwheat (←), *Eriogonum umbellatum*, is native to the western United States.

Rhubarb, *Rheum* x *hybridum,* flowers (←). The fruits (→) are winged achenes. The leaf petioles of this plant are edible after cooking, but the leaf blades are poisonous.

107

Eudicots

Core eudicots

Caryophyllales Perleb

Portulacaceae Jussieu

Genera include:

Claytonia L.—spring beauty
  *Claytonia megarhiza (A. Gray) Parry ex S. Watson—alpine spring beauty
Lewisia Pursh—lewisia
  *Lewisia rediviva Pursh—bitterroot
Portulaca L.—portulaca
  *Portulaca grandiflora Hooker—moss rose
  *Portulaca oleracea L.—common purslane
  *Portulaca umbraticola Kunth—wingpod purslane

* illustrated

Common purslane (↑), *Portulaca oleracea*, can be a troublesome weed. In contrast, wingpod purslane (↓), *P. umbraticola*, is grown as an ornamental.

Claytonia megarhiza (↑), alpine spring beauty, grows in harsh, high altitude environments. Bitterroot (↓), *Lewisia rediviva*, is Montana's state flower.

Members of Portulacaceae are succulents whose leaves are modified for high light intensity and dry environments. The flowers usually have four to six tepals, although some may have more. There are two sepal-like bracts beneath the flower. The pistil is made of two to nine carpels fused together. There are as many branches of the style as there are carpels. Botanists do not have strong evidence that this family holds the descendants of a common ancestor, so its members may be reclassified in the future.

(↑) Moss rose, *Portulaca grandiflora*, has been bred to have extra tepals, and when this occurs, it is called a double flower. The original form had a single row of tepals. The flowers can be white, pink, orange, or red.

Eudicots

Core eudicots

Santalales Dumortier

Santalaceae R. Brown   44 genera/935 species

Genera include:

Arceuthobium M. Bieberstein—western dwarf mistletoe

> **Arceuthobium vaginatum* (Willdenow) J. Presl—pineland dwarf mistletoe, southwestern dwarf mistletoe

Phoradendron Nuttall—mistletoe

> **Phoradendron californicum* Nuttall—desert mistletoe, mesquite mistletoe

> **Phoradendron macrophyllum* (Engelmann) Cockerell—Colorado desert mistletoe, bigleaf mistletoe, broadleaf mistletoe

Santalum L.—sandalwood

NOTE: The former mistletoe family, Viscaceae, is now included in Santalaceae.

* illustrated

Introducing Santalales

Santalales, the mistletoe and sandalwood order, is a small branch of the core eudicots that has an intriguing characteristic. Most species are hemiparasites, which are plants that have chlorophyll, but which also tap into a host plant for water and minerals. Some look like normal plants. They invade the roots of a nearby host plant, so their connections are difficult to find. Others are epiphytes that grow on the branches of a host tree or shrub. The epiphytes have no normal roots and use specialized roots called haustoria to absorb water and nutrients from the branches of their host.

Parasites are often difficult to classify because they lose structures that they no longer use. For example, the flowers of western dwarf mistletoes have lost much of their perianth. The simplified structures are described as reduced. Parasites can become so reduced that it is hard to tell what other plants might be their relatives. DNA data is especially valuable for classifying them.

Not all members of the Santalales have reduced flowers. Members of the tropical family Loranthaceae have showy flowers and are bird-pollinated.

There isn't enough data presently to show where Santalales fits in the core eudicots. Some studies have weak evidence that this order may be a sister to Caryophyllales and the asterids.

Santalaceae, The Mistletoe and Sandalwood Family

Santalaceae is a family of hemiparasites, plants that can make their own food, but which also take water and minerals from other plants. It includes several mistletoes and the sandalwood trees, genus *Santalum*. There are two genera native to the United States, *Phoradendron* and *Arceuthobium*.

The mistletoes grow on the branches of a host tree. They have special roots, called haustoria, that penetrate the host and link to its vascular system. The flowers are small and inconspicuous. The fruits have a sticky outer covering. Some mistletoes shoot their seeds onto the bark of nearby trees. Bird's beaks, feet, and feces carry others to new hosts.

Bigleaf mistletoe (↑), *Phoradendron macrophyllum*, has green leaves and stems. This one is parasitizing a cottonwood tree. Its fruits are white berries that are poisonous to humans, but not to birds.

Desert mistletoe (←), *Phoradendron californicum*, grows on mesquite and other legume trees. Its fruits (↑) are pale red berries. This mistletoe has a partnership with the Phainopepla, a bird that depends on its berries as a major food source and disperses the seeds.

Dwarf mistletoe, *Arceuthobium vaginatum*, lives on ponderosa pine trees. It has brittle, joined stems and varies from golden yellow (↓) to dark red-brown. The host tree, which has gray bark, is usually distorted and stunted by the mistletoe.

The staminate (→) and pistillate (↓) flowers are borne on separate plants, and are very reduced. The mature fruits shoot out their seeds, which may travel as far as 10 meters.

Eudicots

Core eudicots

Saxifragales Dumortier

Crassulaceae J. Saint-Hilaire   34 genera/1370 species

Genera include:

Aeonium Webb & Berthelot—aeonium, Canary Island rose, pinwheel

Crassula L.—crassula

   *Crassula ovata (P. Miller) Druce—jade plant

Dudleya Britton & Rose—cliff rose

Echeveria DC.

   *Echeveria glauca (Baker) E. Morren—hen and chicks

Kalanchoe Adanson—kalanchoe, felt plant, panda plant

   **Kalanchoe daigremontiana* Hamet & Perrier—maternity plant

Sedum L.—sedum, stonecrop, queen's crown, rosecrown

   **Sedum lanceolatum* Torrey—spearleaf stonecrop

Sempervivum L.

   **Sempervivum tectorum* L.—hen and chickens, houseleek

* illustrated

Introducing Saxifragales

The order Saxifragales is a new group in the Angiosperm Phylogeny Group scheme; it is not found in previous classifications. This branch of the core eudicots was proposed after DNA analysis revealed the relationships of its members. Members of this order have a long history, however, with fossils dating back to early branching of the eudicots. There are a wide variety of floral forms in this order, but the flowers usually have separate carpels or carpels that are united only at their bases. Most have a hypanthium, a floral tube or disk, at the base of the flower. The plant forms of this order include trees, shrubs, vines, herbaceous plants, aquatics, and succulents.

This tour visits the stonecrop, gooseberry, peony, and saxifrage families. Other notable members of this order include sweetgum trees (*Liquidambar*), witch-hazel (*Hamamelis*), sweet spire (*Itea*), and water milfoil (*Myriophyllum*).

Just where Saxifragales fits in the core eudicots is not completely clear. Some studies suggest it is a basal group to the rest of the core eudicots, but others put it as a sister group only to the rosids. No study has yet placed this order firmly, so we look forward to further investigations.

Crassulaceae, The Stonecrop Family

Spearleaf stonecrop (↖, ↑), *Sedum lanceolatum,* a native of the western United States, is able to grow on rocky outcrops.

Hen and chicks, *Echeveria glauca* (↑), growing from a soil pocket in a wall

The maternity plant, *Kalanchoe daigremontiana* (↓), forms little plantlets on the margins of its leaves. The plantlets drop off when they reach a certain size, and if they reach the soil, they can root and form new plants.

Jade plant, *Crassula ovata* (↖, ←), has thick, succulent leaves and flowers with five petals, five stamens, and five carpels.

It is difficult to see the venation of stonecrop family leaves, because the leaves are so thick and fleshy. The flowers have equal numbers of sepals, petals, and carpels. The carpels are separate or slightly fused at their bases. In most species, each carpel develops into a follicle. The stems and leaves of these succulent plants root easily if they are broken off and come in contact with the soil. Although many are tender plants found in mild climates, some survive harsh winters and live in alpine areas.

Hen and chickens, *Sempervivum tectorum,* is named for the offshoots the rosettes produce (→). Its genus name means "always living," and refers to its vegetative reproduction. At their center, the flowers have a ring of green carpels that are fused at their bases (←). Two rings of stamens surround the carpels. Once a rosette has bloomed, it dies, but its offshoots live on.

Eudicots

Core eudicots

Saxifragales Dumortier

Grossulariaceae DC.   1 genus/150 species

Genus:

Ribes L.—currant, gooseberry
- **Ribes cereum* Douglas—wax currant
- **Ribes coloradense* Coville—Colorado currant
- **Ribes odoratum* H. L. Wendland—clove currant, golden currant, buffalo currant

* illustrated

Grossulariaceae, The Gooseberry Family

The gooseberries are a family of shrubs. The leaves have several palmate lobes, which makes them easy to recognize. They are spring bloomers with tubular flowers, but the tube is not just the corolla. The calyx is colored and forms part of the tube, which is called a hypanthium or floral cup. Near the end of the hypanthium, the ends of the five sepals flare out. Five small petals attach to the rim of the tube within the whorl of sepals. The nectar is a favorite food source for hummingbirds and bees. The fruits are berries, which are enjoyed by birds and other animals. There are many domesticated varieties of gooseberries and currants.

Ribes species can serve as the alternate host for the white pine blister rust fungus. The primary hosts for this destructive disease are five-needled white pines. This rust fungus must reproduce alternately in the two hosts. In some states, it is illegal to plant *Ribes nigrum*, the European black currant, or other species that are highly susceptible to the fungus. In an effort to control blister rust fungus, over 14 million *Ribes* plants were removed from western national parks between 1930 and 1970. The program was abandoned in 1970 because it did not prove to be effective in protecting pines.

Golden or clove currant (↑), *Ribes odoratum*, has small petals that are edged in red (arrow). The larger yellow lobes are the ends of the sepals. The long yellow tube is the hypanthium or floral tube, a structure that contains fused petal and sepal bases.

Colorado currant (↑), *Ribes coloradense*, has a broad, flat floral cup with tiny petals (arrow) attached between the larger, pink sepal lobes. The five stamens have light-colored anthers.

Wax currant (↓), *Ribes cereum*, lives throughout the western United States. Its pink flowers bloom in early spring and attract hummingbirds. Its inferior ovary (↓) develops into a berry-like fruit that turns bright red in the fall. The dried floral tube still clings to the end of the fruit when it is ripe (↓).

Eudicots

Core eudicots

Saxifragales Dumortier

Paeoniaceae Rafinesque   1 genus/33 species

Genus:

Paeonia L.—peony

    Paeonia lactifolia Pallas—Chinese peony

* illustrated

Paeoniaceae, The Peony Family

Peony buds have leafy bracts beneath them (↑). The sepals have thin, pointed appendages that vary in length (arrow). Sepal size ranges from large, covering half the bud, to small and easy to overlook.

The leaves of peonies are compound, usually with three leaflets, and often lobed. The flowers usually have five sepals of unequal sizes, which are persistent. There are five to nine petals, numerous stamens, and two to five separate carpels (↗). The carpels are attached to a disk at the base of the flower. There is no style; the stigma is shaped somewhat like a hooked, flattened paddle (→). As the flower blooms, the sepals fold down around the outside of the disk. The fruits are leathery follicles (↘). The follicle splits at maturity and reveals the seeds. The seeds that are fertile change from red to black as they mature (lower right). The many infertile seeds remain red and may serve to attract birds, which disperse the seeds. The large-flowered peonies shown here are cultivars of the Chinese peony, *Paeonia lactifolia*.

Double peonies are hybrids whose stamens are converted to petal-like staminodes, which are structures derived from stamens, but which carry no pollen (↓). The many ruffled "petals" in the center of this flower are staminodes.

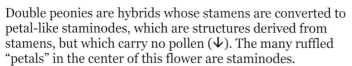

Eudicots

Core eudicots

Saxifragales Dumortier

Saxifragaceae Jussieu   29 genera/630 species

Genera include:

Astilbe Buchanan-Hamilton—astilbe, meadow sweet, false spirea

Bergenia Moench—bergenia

    **Bergenia cordifolia* Sternberg—bergenia

Boykinia Nuttall—boykinia, brook foam

Chrysosplenium L.—golden carpet, golden saxifrage

Heuchera L.—coral bells, alumroot

    **Heuchera parvifolia* Nuttall ex Torrey & A. Gray—alpine alumroot, littleleaf alumroot

    **Heuchera sanguinea* Engelmann—coral bells

Mitella L.—miterwort

*Saxifraga L.—saxifrage, strawberry begonia, rock foil, mossy saxifrage

    **Saxifraga rhomboidea* Greene—snowball saxifrage, diamondleaf saxifrage

Tiarella L.—foamflower

Tolmiea Torrey & A. Gray—piggy-back plant

* illustrated

Saxifragaceae, The Saxifrage Family

There are two separate styles and stigmas in the flowers of bergenia (←), *Bergenia cordifolia*, and mossy saxifrages (→), *Saxifraga* hybrids. A ring of stamens surrounds the pistil.

Alpine alumroot (←, ↑), *Heuchera parvifolia*, lives on cliffs and rock outcrops in the Rocky Mountains.

The saxifrage family are herbs. Many members have a basal rosette of leaves. The leaves of most species are toothed, lobed, or both. The flowers form on a tall, leafless shoot known as a scape. Their sepals are fused at the base, but their five petals are distinct, provided they have petals; some species lack them. There are two or three carpels in the pistil. Several genera of this family have two carpels with the ovaries fused, but have two separate styles. "Saxifrage" means "rock breaker," and these plants often grow in rocky, mountainous areas. They are popular plants for rock gardens.

Snowball saxifrage (↓), *Saxifraga rhomboidea*, grows throughout the Rocky Mountains.

Coral bells (↓), *Heuchera sanguinea*, is a native of Arizona and New Mexico that has been bred as a garden plant. Native *Heuchera* species occur in most of temperate North America.

119

Tree Diagram of Flowering Plant Orders: Rosids

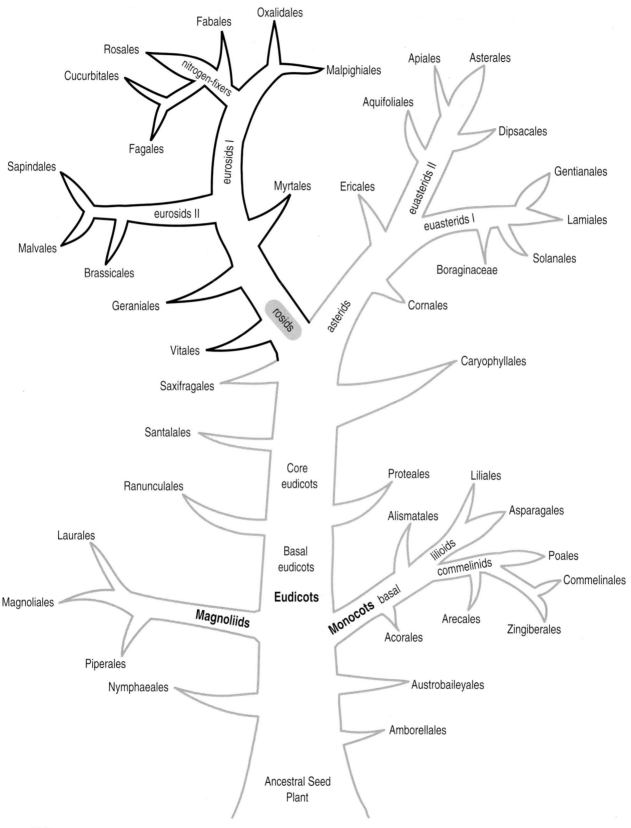

Introducing the Rosids

The rosids make up about a third of flowering plants, but it is not easy to list structures that define them all. As a group, rosids are quite old, with fossils of at least 94 million years of age. They have apparently radiated quickly and developed a wide variety of forms. This makes it difficult to use plant structures to determine their relationships. The DNA tells a great deal of their story, and botanists have made important progress in reading gene sequences and finding the related groups. The picture is still emerging, and the work continues.

Rosids include many woody plants, but also many herbs. Many have numerous stamens, often in two or more whorls. Most have separate petals rather than fused corollas, with the squash family being a notable exception. A number of families, best exemplified by the rose family, have a hypanthium, also called a floral tube or floral cup. The hypanthium covers an inferior or half-inferior ovary and may extend up from it, surrounding the style. The evening primrose family, Onagraceae, has amazing examples of long floral tubes.

Compound leaves are common in rosids, as are stipules, which are usually small, paired structures at the base of the petiole (stem of the leaf). Stipules can be tendrils, spines, leaf-like, or glands.

Plant forms include trees, shrubs, vines, and herbaceous plants. Succulents occur in the euphorbia family. A few members of the loosestrife family and mustard family are aquatics. One family in the oxalis order has a single insectivorous plant, the Australian pitcher plant, *Cephalotus*.

The rosids are divided into two main subgroups, eurosids I and II, as well as several orders of rosids that belong to neither subgroup. Ungrouped rosids on this tour are the grape order (Vitales), the geranium order (Geraniales), and the order of evening primrose, myrtle, and loosestrife families (Myrtales).

The eurosids I are sometimes called the fabids. Eurosid I members that this tour visits are: the squash and begonia families of Cucurbitales; the bean family of Fabales; the alder, oak, and walnut families of Fagales; the diverse order Malpighiales, which includes the willow, violet, flax, spurge, and passionflower families; the oxalis family of Oxalidales; and Rosales, which includes the rose, elm, buckthorn, nettle, and mulberry families.

Eurosids II, sometimes called the malvids, is a smaller rosid branch. This tour visits the mustard family of Brassicales, the mallow family of Malvales, and Sapindales, which includes the sumac, citrus, and soapberry plus maple families.

Eudicots

Rosids

Vitales Reveal

Vitaceae A. Jussieu   14 genera/850 species

Genera include:

Ampelopsis Michaux—blueberry climber, porcelain berry, peppervine

   *Ampelopsis brevipedunculata (Maximowicz) Trautvetter—porcelain berry

Cissus L.—grape ivy, kangaroo vine

Parthenocissus Planchon—wood vine, Boston ivy

   *Parthenocissus quinquefolia (L.) Planchon—Virginia creeper

Vitis L.—grape

   *Vitis labrusca L.—fox grape

* illustrated

Vitaceae, The Grape Family

The fox grape, *Vitis labrusca*, in flower (↑) and with developing fruits (↓). This species is a native of the eastern United States.

The grapes are a family of vines. They climb using tendrils that grow opposite a leaf (↑). The leaves are palmately veined or lobed, or palmately compound. The flowers are tiny and have inconspicuous sepals and petals. Two carpels make up the pistil. The fruit is a berry with one to four seeds.

Porcelain berry (←), *Ampelopsis brevipedunculata*, has tiny flowers with an inconspicuous perianth, but they attract large numbers of bees. The variegated variety, shown here in flower, is less invasive than the green-leaved species and is more desirable as a landscape plant.

Virginia creeper (↓), *Parthenocissus quinquefolia*, has climbing tendrils with adhesive disks at their tips. This differentiates it from *Ampelopsis*, which lacks the disks. Dependable fall color (↓) makes Virginia creeper valuable as an ornamental. (↓) The flowers are inconspicuous.

Eudicots

Rosids

Geraniales Dumortier

Geraniaceae Jussieu   7 genera/805 species

Genera include:

Erodium L'Héritier ex Aiton—storksbill, heronsbill

    **Erodium cicutarium* (L.) L'Héritier ex Aiton—filaree, redstem stork's bill

Geranium L.—wild geranium, hardy geranium, cranesbill

    **Geranium caespitosum* James var. *fremontii* (Torrey ex A. Gray) Dorn—Fremont's geranium

*Pelargonium L'Héritier ex Aiton—garden geranium, scented geranium

    *Pelargonium x *hortorum* L. H. Bailey—garden geranium, zonal geranium

* illustrated

The geranium family grows worldwide, and many members are used as ornamental plants. The North American natives have flowers with five identical petals. Our imports include common garden geraniums, genus *Pelargonium*, which have two petals that differ slightly from the rest. The stamens mature first, and then the five stigmas open and the anthers fall off. (See lower right photo.) The sepals persist after the flower blooms. The fruit is long and thin, which has inspired the common names of cranesbill and storksbill. Five carpels make up the pistil. The styles surround a central column, around which the fruits develop. The leaves of this family are palmately veined and are often lobed. They have hairs with aromatic oils. The scented geraniums are named for the fragrance of their leaves; some smell like roses, citrus fruit, mints, and other quite unrelated plants.

(↑) Filaree or storksbill, *Erodium cicutarium*, is an early spring bloomer throughout North America. This small weedy plant is named for its fruits. Its genus name comes from the Greek *erodios*, meaning "heron." The fruit is a capsule that splits into five sections. Each seed has a corkscrew awn (↑) that helps push it into the soil.

Fremont's geranium (↑), *Geranium caespitosum* var. *fremontii*, is native to the southern Rocky Mountains. It is a hardy perennial. The seeds develop (↑) under the cover of the persistent calyx (white arrow). Its fruits are capsules that split and coil up when they are mature (↑), flinging out their seeds.

Garden and houseplant geraniums (↓) belong to genus *Pelargonium*, as do scented geraniums (↓), ivy geraniums, and regal geraniums. This genus originated in South Africa.

Eudicots

Rosids

Geraniales

125

Eudicots

Rosids

Myrtales Reichenbach

Lythraceae J. Saint-Hilaire   31 genera/620 species

Genera include:

Cuphea P. Browne—cuphea, cigar plant
      *Cuphea llavea Lexarza—bat-faced cuphea
Decodon J. Gmelin—swamp loosestrife
Lagerstroemia L.
      *Lagerstroemia indica L.—crepe myrtle
Lawsonia L.—henna, mignonette tree
Lythrum L.—purple loosestrife
Punica L.
      *Punica granatum L.—pomegranate

* illustrated

The flowers of Lythraceae have a calyx tube that persists after bloom and expands as it covers the developing fruit. The flower petals are crumpled in the bud and remain wrinkled after they open. The stamens attach inside the calyx tube and have filaments of different lengths. The pistil has two or more carpels, but one style. The fruits of this family are usually capsules, but include berries like the pomegranate as well.

The pomegranate (↖, ←), *Punica granatum*, is native to western Asia and has been cultivated in the Mediterranean region since prehistoric times. The fruit of pomegranates is a large berry whose seeds have a fleshy outer covering. This juicy seed covering is the edible part. The flower's calyx tube becomes the fruit's thick, leathery rind. The carpel walls enclose the seeds in several chambers.

The genus *Cuphea* has tubular flowers. The fused sepals form the tube. Bat-faced cuphea (→), *C. llavea*, has only two petals, which are scarlet. They attach at the end of the calyx tube.

Crepe myrtle (↑), *Lagerstroemia indica*, has strongly clawed petals, which means that the petals taper to a narrow base. Its calyx tube has pointed lobes. (↑) After the flower blooms, the points fold inward and cover the developing ovary. As the ovary grows, it protrudes out of the persistent calyx (↑).

Eudicots

Rosids

Myrtales Reichenbach

Myrtaceae Jussieu   131 genera/4620 species

Genera include:

Acca O. Berg

    *Acca sellowiana (O. Berg) Burret—pineapple guava, feijoa

*Callistemon R. Brown—bottlebrush

Chamelaucium Desfontaines—florist's waxflower

*Eucalyptus L'Héritier—eucalyptus tree, gum tree

Eugenia L.—Surinam cherry, Australian bush cherry

Leptospermum J. R. Forster & G. Forster—tea tree

    **Leptospermum scoparium* J. R. Forster & G. Forster—tea tree

Metrosideros Banks ex Gaertner—New Zealand Christmas tree

Myrtus L.—myrtle

Pimenta Lindley—allspice, pimento, bay rum tree

Psidium L.—guava, strawberry guava

Syzygium Gaertner—clove tree, Malay apple, rose-apple

* illustrated

Myrtaceae, The Myrtle Family

Prominent stamens characterize many flowers of the family Myrtaceae. In several genera, the stamens are colored and perform the function of attracting pollinators. The flowers have a broad disk that covers the inferior ovary. The stamens attach to the edge of this disk. The leaves are dotted with little pockets of aromatic compounds. This family grows in tropical and subtropical areas. It has many members in Australia, several of which are grown as landscape plants in the United States. In California, a number of eucalyptus species have naturalized.

Flowers of pineapple guava (↑), *Acca sellowiana*, previously classified as *Feijoa sellowiana*. The plants and the fruits of this South American native are commonly called feijoas.

Eucalyptus flowers (↑) have a thick fringe of stamens. The sepals and petals form a cap that covers the stamens in bud and falls off when the flower blooms. Despite the numerous stamens, there is one style and one stigma. The fruits (→) are woody and have an X-shaped opening from which the seeds exit.

(↑) Stamens of bottle-brush flowers, *Callistemon*, have prominent red filaments. Its developing fruits (←) retain the style and stigma at first, as well as the floral tube. Later, this tube or hypanthium becomes a woody covering around the fruit.

A pink-flowered cultivar of tea tree (↑), *Leptospermum scoparium* shows the broad floral disk with the red stamens attached at its rim.

129

Eudicots

Rosids

Myrtales Reichenbach

Onagraceae Jussieu   17 genera/650 species

Genera include:

Boisduvalia Spach—spike primrose

Camissonia Link—suncup, evening primrose

Circaea L.—enchanter's nightshade

Clarkia Pursh—clarkia, godetia, farewell-to-spring

Epilobium L.—fireweed, willowherb

    **Epilobium angustifolium* L.—fireweed, great willowherb

    *Epilobium canum (Greene) P. H. Raven—California fuchsia, zauschneria

Fuchsia L.—fuchsia

    *Fuchsia x *hybrida* Horticulture ex Siebold & Voss—hybrid fuchsia

Gaura L.—gaura

Ludwigia L.—seedbox

Oenothera L.—evening primrose

    **Oenothera biennis* L.—common evening primrose

    **Oenothera macrocarpa* Nuttall—Ozark sundrops, Missouri primrose, bigfruit evening primrose

    **Oenothera speciosa* Nuttall—showy evening primrose, Mexican evening primrose

Zauschneria C. Presl is now a part of *Epilobium* L.

* illustrated

Onagraceae, The Evening Primrose Family

Flowers of Onagraceae have four sepals, four petals, four to eight stamens, and a four-carpellate pistil. The pollen is bound together by sticky threads. Strings of pollen can be seen on the anthers of evening primroses without using a magnifier (↓). The stigma has four lobes that may be short or long. The ovary is inferior (white arrows in the photos below). A floral tube connects the ovary to the base of the perianth. In some species, this tube is much longer than the corolla itself. The fruit is usually a capsule that splits into four sections, each derived from a carpel.

Showy evening primrose (↑), *Oenothera speciosa*

California fuchsia, *Epilobium canum* (↑), formerly *Zauschneria californica*
A hybrid fuchsia, *Fuchsia* x *hybrida*, with pink sepals and purple petals (↑)

Fireweed, *Epilobium angustifolium*, in bloom and with mature, splitting capsules (→). The small seeds are wind-dispersed. Fireweed is known for growing in disturbed soils such as burned areas.

Common evening primrose, *Oenothera biennis,* inflorescence (←) with buds at the top and developing ovaries below. The mature capsules (↓) split in four parts.

Ozark sundrops or Missouri primrose (→), *Oenothera macrocarpa*, has a very long floral tube. The bracket indicates the floral tube. The inferior ovaries develop into a fruit with four wings. Two fruits can be seen in the foreground (arrows).

Eudicots

Eurosids I

Cucurbitales Dumortier

     Begoniaceae Berchtold & J. Presl   2 genera/1400 species

        Genera:
        Begonia L.—begonia, including fibrous, tuberous, and Rex begonia
            Begonia masoniana Irmscher—iron cross begonia
        Hillebrandia Oliver—hillebrandia, aka'aka'aka

* illustrated

The Branch of the Nitrogen-Fixing Partnerships

It appears that sometime back in the reaches of angiosperm history, an ancestral rosid developed the ability to form a beneficial symbiotic relationship with nitrogen-fixing bacteria. Among its descendants are all the plants that are able to house nitrogen-fixing bacteria in root nodules. This clade holds the orders Fabales, Fagales, Rosales, and Cucurbitales.

Members of these orders that have the nitrogen-fixing symbiosis were formerly classified in several unrelated families and orders. It is hard to see how such a complex process could have arisen several times. Root nodule formation requires many steps and many gene products from both the host and the bacterium. Phylogenic classification gives a more logical picture, since it is more probable that all these families shared a common ancestor.

The most famous root nodule formation occurs between the Fabaceae, the bean or legume family, and bacteria of genus *Rhizobium* and related genera. This symbiosis allows the plants of Fabaceae to grow in poor soils. The plants not only thrive, they add nitrogen to the soil. Before the advent of the Haber process for chemical nitrogen fixation, legume crops and their nitrogen-fixing bacteria offered agriculture a vital means to increase soil fertility.

In other families of this clade, the filamentous bacteria of genus *Frankia* form nitrogen-fixing root nodules. In the Rosales, nitrogen fixers in the rose family, Rosaceae, include the alpine dryads of genus *Dryas*, mountain mahogany of genus *Cercocarpus*, fernbush of genus *Chamaebatiaria*, and cliffrose of genus *Purshia*. The buckthorn family, Rhamnaceae, has several nitrogen-fixing genera, including *Ceanothus*, which includes common shrubs of the California chaparral biome. Buffalo berry of genus *Shepherdia* and Russian olive of genus *Elaeagnus* are nitrogen-fixers of the family Elaeagnaceae.

In the Cucurbitales order, there are two families that have nitrogen-fixing members. Coriariaceae is a family of shrubs whose native habitats are scattered in several continents—Central and South America, the Mediterranean area, and parts of Asia and the South Pacific. Some members of its sole genus, *Coriaria*, are grown as ornamentals. Datiscaceae is a small family with a single genus of herbaceous perennials, *Datisca*. There are species native to California and Mexico, as well as the Middle East and northern India.

Biogeography certainly didn't unite the nitrogen-fixing species, nor did structural studies. The story was revealed by their DNA sequences.

Begoniaceae, The Begonia Family

(←) Staminate flowers have yellow anthers.

A pistillate flower (→) shows its twisted yellow stigmas.

The begonias have leaves with an asymmetrical base. The leaves are thick and nearly succulent, and they are often toothed or lobed. The plants are monoecious. The staminate flowers usually have two large and two small tepals. The pistillate flowers usually have five tepals. Their ovary is inferior and has three wings (→). The stigmas of the three-carpellate pistil are curled and are bright yellow, just like the anthers. This coloring tricks pollen-carrying insects into landing on the stigma in search of more pollen, and so promotes pollination.

The pistillate flower of this tuberous begonia has a green, winged ovary (→).

Begonias are native to tropical climates. Begonias in temperate North America are ornamentals, and usually hybrids. They are cultivated as houseplants and grown as summer annuals. Many begonias, such as the three shown below, are grown for their striking foliage.

Iron cross (↓), *Begonia masoniana*

Eudicots

Eurosids I

Cucurbitales Dumortier

Cucurbitaceae Jussieu   118 genera/845 species

Genera include:

Apodanthera Arnott—melon-loco

Citrullus Schrader ex Echlon & Zeyher—watermelon

Cucumis L.—cucumber, gherkin, cantaloupe, muskmelon, honeydew melon
> *Cucumis sativus* L.—cucumber

Cucurbita L.—summer and winter squash, marrow, pumpkin, ornamental gourd
> *Cucurbita pepo* L.—summer squash

Ecballium A. Richard—squirting cucumber

Echinocystis Torrey & A. Gray—wild cucumber, bur cucumber, wild balsam-apple

Lagenaria Seringe—bottle gourd, calabash

Luffa Miller—loofa gourd, vegetable sponge

Marah Kellogg—manroot, wild cucumber, spiny cucumber
> *Marah fabaceus* (Naudin) Naudin ex Greene—California manroot
> *Marah oreganus* (Torrey ex S. Watson) T. J. Howell—coastal manroot

Sechium P. Browne—chayote

Sicyos L.—wild cucumber, star cucumber

* illustrated

The squashes are important food plants. They are bush-like herbs or vines with tendrils. The leaves are palmately veined and have coarse, rough hairs. The plants are monoecious. The pistillate flowers have an inferior ovary. This is the "baby squash" beneath the corolla (↓). The three lobes of the stigma and the three sections of the fruits reflect the three-carpellate pistil. All the stamens are united into a single structure. The five petals (←) are fused at their bases, which is unusual for a rosid.

Many varieties of summer squashes, *Cucurbita pepo*, exist. They include zucchini, golden zucchini (↓), yellow crookneck, and patty pan.

The pistillate flower (↑) has three stigma lobes. The corolla sits atop the yellow inferior ovary (→).

Staminate flowers have all the stamens fused together (↑) and a plain green stem (←) .

Along the California coast, several species of genus *Marah* grow. The California manroot (↓), *M. fabaceus*, has a spherical fruit that is covered in prickles. The fruits of coastal manroot (↙), *M. oreganus*, have fewer prickles and are slightly beaked. The prickles are present on the ovary of the pistillate flower (←).

This is the pistillate flower (↑) of a parthenocarpic or greenhouse cucumber, a horticultural variety of *Cucumis sativus*. It develops fruits without pollination, and therefore has seedless fruits. Pistillate flowers are always formed, but staminate flowers form only during periods of stress or if plant hormones are applied.

Eudicots

Eurosids I

Fabales Bromhead

Fabaceae Lindley or Leguminosae Jussieu   750 genera/19400 species
Subfamily Faboideae or Papilionoideae

Genera include:

Amorpha L.—leadplant

Arachis L.—peanut

Astragalus L.—locoweed, milk vetch

Baptisia Ventenat—wild indigo

Cytisus Desfontaines—Scotch broom

Dalea L.—dalea, prairie clover

Erythrina L.—coral tree
> *Erythrina fusca Loureiro—bucayo, kaffirboom coral tree

Genista L.—broom

Glycine Willdenow—soybean

Glycyrrhiza L.—licorice

Indigofera L.—indigo

Lablab Adanson
> *Lablab purpureus (L.) Sweet—hyacinth bean

Lathyrus L.—sweet pea

Lens Miller—lentil

*Lupinus L.—lupine

Medicago L.—alfalfa, lucerne, medick, bur clover
> *Medicago sativa L.—alfalfa

Melilotus Miller—sweet clover, yellow clover

Olneya A. Gray—ironwood

Oxytropis DC.—locoweed
> *Oxytrophis sericea Nuttall—white locoweed

Phaseolus L.—bean

Pisum L.—pea

Robinia L.—locust tree, black locust
> *Robinia neomexicana A. Gray—New Mexico locust

Sophora L.—necklace pod, mescal bean tree, sophora, Japanese scholar tree

Spartium L.—Spanish broom

Thermopsis R. Brown—golden pea
> *Thermopsis montana Nuttall—mountain golden pea, false lupine

Trifolium L.—clover
> *Trifolium pretense L.—red clover

Ulex L.—gorse

Vicia L.—vetch

Wisteria Nutall—wisteria

* illustrated

Another Group of Fabaceae
Cercideae was formerly included in the subfamily Caesalpinioideae, but genetic analysis showed that it is an independent branch at the base of the Fabaceae. The redbuds (←), genus *Cercis*, and the orchid trees (→), genus *Bauhinia* are members of this tribe. The leaves are frequently bilobed (→). Redbud flowers look similar to bean flowers, but they have differences in structure.

Fabaceae or Leguminosae, The Bean Family
Faboideae, The Bean Subfamily

(↖) Alfalfa, *Medicago sativa*, a forage crop plant

(↑) Hyacinth bean, *Lablab purpureus*

(↗) Golden pea, *Thermopsis montana*, a native wildflower with poisonous pods

(→) Red clover, *Trifolium pretense*, has trifoliate leaves.

(←) Garden lupine, a *Lupinus* hybrid, has palmately compound leaves.

The bean family includes herbs, shrubs, and trees and has four subgroups. They are known for their root nodules, which contain nitrogen-fixing bacteria. The flowers of the subgroups look very different, but they all have one carpel with a superior ovary. Most have bean pod–like fruits, which are called legumes.

Most flowers of the bean subfamily have a typical sweet pea shape, bilaterally symmetrical with petals of different shapes. The stamens have the base of their filaments fused around the carpel. In some, all but one of the 10 stamens are fused. The leaves are compound in various ways. Members include many important food plants, such as beans, soybeans, lentils, peas, and peanuts.

(↓) White locoweed, *Oxytropis sericea*   Bucayo or kaffirboom coral tree, *Erythrina fusca* (↓)
New Mexico locust (↓), *Robinia neomexicana*

137

Eudicots

Eurosids I

Fabales Bromhead

Fabaceae Lindley or Leguminosae Jussieu

750 genera/19400 species

Subfamily Mimosoideae

Genera include:

Acacia Miller—acacia, wattle tree, bullhorn acacia, catclaw acacia
  *Acacia melanoxylon R. Brown—blackwood acacia
Albizia Durazzini
  *Albizia julibrissin Durazzini—silktree, mimosa
Calliandra Bentham—powderpuff
Mimosa L.—sensitive plant
  *Mimosa rupertiana B. L. Turner—sensitive briar
Prosopis L.—mesquite
  *Prosopis glandulosa Torrey—honey mesquite

* illustrated

Eudicots

Eurosids I

Fabales Bromhead

Fabaceae Lindley or Leguminosae Jussieu

750 genera/19400 species

Subfamily Caesalpinioideae

Genera include:

Caesalpinia L.—bird-of-paradise, holdback, nicker (includes the former *Poinciana*)
  *Caesalpinia pulcherrima (L.) Swartz—Barbados pride, peacock flower,
    red bird-of-paradise
Cassia L.—cassia, golden shower, pink shower
Gleditsia L.—honey locust
Gymnocladus Lamark
  *Gymnocladus dioica (L.) K. Koch—Kentucky coffee tree
Hoffmannseggia Cavanilles—rushpea
  *Hoffmannseggia glauca (Ortega) Eifert—hog potato, mesquite weed
Parkinsonia L.—palo verde, Jerusalem thorn
Senna Miller—senna
  *Senna roemeriana (Scheele) Irwin & Barneby—twoleaf senna
Tamarindus L.—tamarind

Fabaceae or Leguminosae: Mimosoideae, The Mimosa Subfamily

The flower clusters of the mimosa subfamily look like little pom-poms. Their petals are tiny and the stamens are very long. The leaves are usually bipinnately compound.

Sensitive briar (→), *Mimosa rupertiana*, has leaflets that fold up when they are touched.

Silk tree or mimosa (←), *Albizia julibrissin*

Honey mesquite (→), *Prosopis glandulosa*

The leaves of this blackwood acacia tree (←), *Acacia melanoxylon*, appear to be simple, but instead, they are expanded, flattened petioles, which are called phyllodes. The leaf blade does not form.

Fabaceae or Leguminosae: Caesalpinoideae, The Caesalpinia Subfamily

The Caesalpinia subfamily has flowers with five similar petals. The leaves are compound, but may have just two leaflets (↓).

Hog potato, *Hoffmannseggia glauca* (→), is a small, weedy species with bipinnately compound leaves.

(↓) Barbados pride, *Caesalpinia pulcherrima*, is an ornamental shrub grown in areas with mild winters.

Twoleaf senna, *Senna roemeriana* (↓), is native to Texas and Mexico. It is poisonous to livestock.

Staminate flowers of Kentucky coffee tree, *Gymnocladus dioica*, a dioecious species with large, bipinnately compound leaves (↘)

Eudicots

Eurosids I

Fagales Engler

Betulaceae Gray   6 genera/110 species

Genera include:

Alnus Miller—alder

    *Alnus incana (L.) Moench—mountain alder

Betula L.—birch

    *Betula nigra L.—river birch

Carpinus L.—ironwood, hornbeam

Corylus L.—hazelnut, filbert

    *Corylus avellana L.—common filbert

Ostrya Scopoli—hop hornbeam

* illustrated

140

Betulaceae, The Birch and Alder Family

Mountain alder, *Alnus incana*, shown (1) in early spring with the remains of last year's fruits and this year's immature inflorescences (white arrow, pistillate; black arrow, staminate); (2) as the staminate flowers mature; (3) in early fruit development; (4) midsummer green fruits; and (5) in late summer with next year's staminate and pistillate inflorescences (arrows).

The flowers of Betulaceae are unisexual, with both sexes on one plant. The flowers form the summer before their spring bloom. The staminate catkins (1, black arrow) are larger than the pistillate inflorescences (1, white arrow). After they bloom, the staminate catkins fall off and the pistillate inflorescences enlarge and become green, cone-like structures in which the fruits grow. In the fall, the remains of the alder pistillate flowers are dried bracts that look like a tiny pinecone. The bracts of the mature pistillate flowers of birches are shed along with the fruits. This family has doubly serrate leaves.

Staminate and pistillate catkins of the common filbert, *Corylus avellana* Contorta (↑). The pistillate flowers (↑) are visible as tiny, red thread-like stigmas. The pistillate flowers develop into bract-covered nuts.

The river birch, *Betula nigra*, (→) has pistillate inflorescences that stand upright, unlike alder pistillate inflorescences shown in the top photos.

Eudicots

Eurosids I

Fagales

141

Eudicots

Eurosids I

Fagales Engler

Fagaceae Dumortier   7 genera/670 species

Genera include:

Castanea Miller—chestnut, chinkapin

Chrysolepis Hjelmquist—giant chinkapin, bush chinkapin

Fagus L.—beech

Fagus sylvatica L.—European beech

Lithocarpus Blume—tanoak

Quercus L.—oak

Quercus rubra L.—northern red oak

* illustrated

(↑) The pistillate flowers of oaks are tiny and inconspicuous. They are three-carpellate, with an inferior ovary that is covered in bracts. There is no perianth. The staminate flowers are in catkins (↑), as they are in all the families of this order. The fruits of oaks are acorns. The cap is several rows of bracts that surround the developing fruit (↓). The shapes of the simple, alternate leaves range from ovate with entire margins to the familiar lobed oak leaves. (↑) Northern red oak, *Quercus rubra*, in bloom

(←) As the young acorns develop, they are almost covered by the bracts of their cup at first. The remains of the three styles still protrude from the bracts. Later, the green acorn grows out of the bracts, still bearing the remains of the tiny styles and stigmas.

The acorns take one or two years to mature, depending on the species of oak.

The buds of oaks (↑), genus *Quercus*, and beeches (↑), genus *Fagus*, are covered in a beautiful pattern of overlapping bud scales. Shoots of European beech (→), *Fagus sylvatica*, emerge from their pointed bud scales in May.

Eudicots

Eurosids I

Fagales

143

Eudicots

Eurosids I

Fagales Engler

Juglandaceae Perleb   7 genera/50 species

Genera include:

Carya Nuttall—pecan, hickory
    *Carya illinoensis (Wangenheim) K. Koch—pecan
Juglans L.—walnut, butternut
    *Juglans microcarpa Berlandier—little walnut, river walnut
Pterocarya Kunth—wingnut

* illustrated

Flowers of Juglandaceae are unisexual, with both sexes on one tree. The leaves are odd-pinnately compound. The staminate flowers form in long catkins. The pistillate flowers (←, ↘) grow into green drupes that have a leathery outer husk. Inside the husk, a stony endocarp, like a peach pit, encloses the seed. Botanically speaking, what we crack is not a nut, but rather this endocarp.

Little walnut (←), *Juglans microcarpa*, (1) pistillate flowers; (2) young drupes; (3) developing drupes; and (4) mature drupe that has dried and cracked, revealing the brown, ridged endocarp that encloses the seed.

Pecan trees, *Carya illinoensis*, have staminate catkins (↑) with three branches. The ovary of the pistillate (↑) flower develops into a green drupe with four seams. In the fall, drupes of genus *Carya* split (↓) and drop their endocarps. These are dried and aged before they are cracked to harvest the seed, which is commonly called pecan nutmeats or pecan halves. In walnuts, genus *Juglans*, the drupes do not split apart.

The pecan endocarp with its enclosed seed (→) lies inside the outer husk, which becomes dry and brittle as the fruit matures.

Eudicots

Eurosids I

Fagales

145

Eudicots

Eurosids I

Rosales Perleb

Moraceae Link   38 genera/1100 species

Genera include:

Artocarpus J. R. Forster & G. Forster—breadfruit

Broussonetia L'Héritier ex Ventenat—paper mulberry

Ficus L.—fig, strangler fig, weeping fig, rubber tree, bo tree, banyan tree

Ficus carica L.—common fig

Maclura Nuttall

Maclura pomifera (Rafinesque) C. Schneider—Osage orange, bois d'arc

Morus L.—mulberry

Morus nigra L.—black mulberry

* illustrated

Moraceae, The Fig and Mulberry Family

Common fig (←), *Ficus carica*

Osage orange, bois d'arc (→), *Maclura pomifera*

The diverse members of the Moraceae commonly have multiple fruits, also called syncarps. These fruits are formed as the ovaries of many separate flowers merge. An enlarged receptacle common to all the flowers or an enlarged, fleshy perianth often join the individual ovaries together. All family members have milky sap.

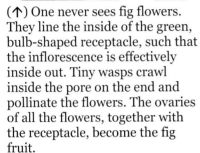

(↑) One never sees fig flowers. They line the inside of the green, bulb-shaped receptacle, such that the inflorescence is effectively inside out. Tiny wasps crawl inside the pore on the end and pollinate the flowers. The ovaries of all the flowers, together with the receptacle, become the fig fruit.

(↑) A cross section of an Osage orange reveals the seeds, each of which was formed from the ovary of a separate flower.

Inflorescences of black mulberry (↑), *Morus nigra*, appear as soon as the buds unfurl. The flowers are unisexual. The whole pistillate inflorescence (↑) will become a single mulberry. The developing fruits still have the remains of the two stigmas (↑) on each segment. The ripe fruit (→) is an aggregate of drupes that are each surrounded by enlarged, fleshy tepals.

Eudicots

Eurosids I

Rosales Perleb

Rhamnaceae Jussieu   52 genera/925 species

Genera include:

Ceanothus L.—California lilac, ceanothus, buckbrush, redroot, New Jersey tea

   Ceanothus fendleri A. Gray—buckbrush, Fendler's ceanothus

Krugiodendron Urban—leadwood

Rhamnus L.—buckthorn, cascara sagrada, redberry, coffeeberry

   Rhamnus cathartica L.—common buckthorn

   Rhamnus frangula L.—alder buckthorn, columnar buckthorn

Ziziphus Miller (*Zizyphus* Adanson)—Chinese jujube

   Ziziphus obtusifolia (Hooker ex Torrey & A. Gray) A. Gray—graythorn,
   lotebush

* illustrated

Rhamnaceae, The Buckthorn Family

The Rhamnaceae are a family of spring-flowering shrubs and trees. Their flowers are small, but often occur in large masses. Four or five tiny, cup-like petals that are often clawed attach to a disk at the base of the flower. A stamen lies on top of each petal. The pistil is usually three-carpellate. Fruits are either a berry-like drupe or one that splits into three sections and flings out its seeds.

Ceanothus is a genus of about 55 species of shrubs, some of which have been hybridized for ornamental use. This genus makes up a significant part of the chaparral biome of California. The shrubs fix nitrogen with the help of the bacteria called *Frankia*. Their fruits are three-part drupes that split apart at maturity.

(←) *Ceanothus* hybrids are sometimes called California lilacs, although they are not related to the true lilacs of the family Oleaceae. Most California native species have blue flowers.

Graythorn (←), *Ziziphus obtusifolia*, is native to the southwestern United States. It is used as an ornamental in arid areas.

One of the many shrubs called buckbrush, *Ceanothus fendleri* (↑) is native to the southern Rocky Mountain area and Texas.

Alder buckthorn (←,↑), *Rhamnus frangula*, is an Old World species that is used as an ornamental. Its fruits are berry-like. The green fruits of common buckthorn, *Rhamnus cathartica*, have the persistent floral disk at their base and the three-fold stigma at their apex (↑). This species is also an Old World native.

Eudicots

Eurosids I

Rosales Perleb

Rosaceae Jussieu   95 genera/2830 species

Genera include:

Alchemia L.—lady's mantle

Amelanchier Medikus—serviceberry

Aronia Medikus—chokeberry

Cercocarpus Kunth—mountain mahogany

Cotoneaster Medikus—cotoneaster

Crataegus L.—hawthorn

Erythrocoma Greene

> **Erythrocoma triflora* (Pursh) Greene—pink plumes, prairie smoke, old man's whiskers

Fallugia Endlicher—Apache plume

Filipendula Miller—meadowsweet, queen-of-the-prairie

Fragaria L.—strawberry

Geum L.—avens, geum

Holodiscus (K. Koch) Maximowicz—ocean spray, rock spirea

Malus Miller—apple, crabapple

Physocarpus (Cambessèdes) Maximowicz—ninebark

Potentilla L.—cinquefoil, potentilla

> **Potentilla fruticosa* L.—shrubby cinquefoil

Prunus L.—almond, apricot, cherry, cherry-laurel, damson, plum, peach, nectarine

Pyracantha M. Roemer—pyracantha

Pyrus L.—pear

*Rosa L.—rose

Rubus L.—blackberry, raspberry, bramble

Sanguisorba L.—salad burnet

Sorbaria (DC.) A. Braun—false spirea

> **Sorbaria sorbifolia* (L.) A. Braun—Ural false spirea

Sorbus L.—mountain ash, whitebeam

Spiraea L.—spirea

* illustrated

Rosaceae, The Rose Family

Roses, genus *Rosa*, are familiar and easily recognized flowers, but many members of this family do not look like cultivated roses. Hybrid tea roses (↑) serve to show the features of the flowers. They typically (↑) have a cup-shaped receptacle (arrow). The five sepals, the numerous stamens, and the petals attach to the rim of this floral cup. The ovaries of the one or more separate carpels are embedded down in the receptacle (↓, arrow). The styles and stigmas stick out from the middle of the flower (↑). In the photo above, the stamens have dark anthers and surround the lighter stigmas of the carpels. A rose hip is the thickened, floral cup that contains mature achenes, which are the true fruits. In the example below, the floral cup is orange (↘) with persistent green sepals. There are many kinds of fruits in this family, some of which are unique to it. The following page has information about domestic fruits from the rose family.

Prairie smoke, also called old man's whiskers, *Erythrocoma triflora* (↑), is named for its feathery fruits. Field guides often call it *Geum triflorum*.

Shrubby cinquefoil, *Potentilla fruticosa*, (↘) is native to a wide area of the western and northern United States. It is probably not closely related to other *Potentilla* members and will be renamed in the future.

False spirea (→), *Sorbaria sorbifolia*, is one of several shrubs with a mass of small flowers. Others include spirea (*Spiraea*) and meadowsweet (*Filipendula*).

Rosaceae species illustrated on the opposite page:

Fragaria x *ananassa* Duchesne ex Rozier—cultivated strawberry
Malus pumila Miller—apple
Prunus armeniaca L.—apricot
Rubus fruticosus L.—shrub blackberry

Domestic Fruits of the Rose Family

The rose family provides many edible fruits. The fruits take such different forms that many people do not realize that come from one family. Botanically, fruits are the mature ovaries. The fleshy part of fruits we eat is generally derived from layers of the ovary wall. Apples and strawberries are exceptions. Their fruits, known botanically as accessory fruits, include extra tissue from the receptacle or floral cup.

Apricots (↑), peaches, nectarines, and cherries have one carpel in their flowers. Their fruits are drupes, commonly called stone fruits. The exocarp and mesocarp, which are the skin and pulp of the fruit, are the edible parts. The stony endocarp has the seed inside. Almonds are also a member of this group. Their outer layers are leathery. The edible part of almonds is the seed within the endocarp.

Blackberries (↑) and strawberries (↓) have many separate carpels in their flowers. In the blackberry, the ovaries become druplets, small drupe-like fruits that aggregate to form the berry. The dried style can be found on the druplets of immature blackberries (↑).

What we call the seeds are really the strawberry fruits (↓). They are embedded on the outside of their enlarged, red fleshy receptacle.

Apple flowers (↑), have three to five carpels. The fleshy, edible part of the fruit forms from the enlarged floral cup. The sepals attach to the rim of the cup, and one can often find their remains on the end of the mature fruit. The ovary becomes the apple's core. The tough membrane around the seeds is the endocarp, the inner wall of the ovary.

153

Eudicots

Eurosids I

Rosales Perleb

Ulmaceae Mirbel   6 genera/35 species

Genera include:

Planera J. F. Gmelin—planer tree, water elm

Ulmus L.—elm tree

   Ulmus americana L.—American elm

   Ulmus pumila L.—Siberian elm

Zelkova Spach—zelkova tree

* illustrated

Ulmaceae, The Elm Family

(↑) Siberian elm, *Ulmus pumila*, blooms when there is still freezing weather. It is wind-pollinated and produces abundant, often allergy-triggering, pollen. The ovaries quickly enlarge, (↑) appearing as green disks among the dried black anthers. The mature fruits (→) blow away before the leaves are large enough to block the wind.

Ulmaceae is a family of trees with tiny, bisexual flowers that have no petals and are wind-pollinated. Their winged, dry fruits, which are known botanically as samaras, mature and are shed as the new foliage begins to grow in the spring. The word samara literally means "elm seed" in Latin.

(↖) The flowers of the American elm, *Ulmus americana*, are borne on short stems. Like other elms, their pistils mature before the stamens. (↑) The first sign of the developing ovary is the two styles growing out of the calyx. The samaras, which are edged with fine white hairs, hang in clusters before the wind disperses them (↑).

(←) The leaf of the American elm is typical of the family—a simple leaf with a serrate margin in an alternate leaf arrangement, with strong pinnate venation and an asymmetrical base.

The American elm population has been decimated by a lethal fungus. Bark beetles carry the fungus from tree to tree. The pathogen was introduced into the United States via contaminated wood from Europe. Although it is called Dutch elm disease, the infestation may have originated in Asia. Asian elms are resistant to it, and have been planted to replace American elms. Plant breeders have produced disease-resistant varieties of American elm that may help re-establish the population.

Eudicots

Eurosids I

Rosales Perleb

Urticaceae Jussieu   54 genera/2625 species

Genera include:

Boehmeria Jacquin—ramie, false nettle

    *Boehmeria biloba Weddell—Japanese false nettle

Laportea Gaudichaud-Beaupré—wood nettle

Pilea Lindley—aluminum plant, artillery plant, clearweed, rock weed

    *Pilea cadierei Gagnepain & Guillaumin—aluminum plant

    *Pilea mollis Weddell—moon valley pilea

    *Pilea nummulariifolia (Swartz) Weddell—creeping charlie

Soleirolia Gaudichaud-Beaupré—baby's tears, mother-of-thousands

Urtica L.—nettle

    *Urtica dioica L.—stinging nettle

* illustrated

Urticaceae, The Nettle Family

The nettle family is best known for the stinging hairs on some members' leaves. Histamine is one of the chemicals in nettles' sting. The minute flowers are unisexual and usually on separate plants. The stamens are folded into the center of the flower until it blooms. They pop out and fling their pollen as the flower opens. The artillery plant is named for the puffs of pollen from the staminate flowers.

Stinging nettles (←, ↑), *Urticaria dioica*, have tiny whitish flowers. The long, narrow leaves are opposite and have toothed margins.

Houseplants from the genus *Pilea* of the nettle family include moon valley pilea, *P. mollis* (↑); creeping charlie, *P. nummulariifolia* (↑); and aluminum plant, *P. cadierei* (↓).

Japanese false nettle, *Boehmeria biloba* (↑), is a relative of ramie, *B. nivea*. Ramie provides an especially strong fiber that is used in clothing.

Eudicots

Eurosids I

Malpighiales Martius

Euphorbiaceae Jussieu   222 genera/5970 species

Genera include:

Acalypha L.—chenille plant, copper leaf, beefsteak plant

Codiaeum A. Jussieu—croton (houseplant)

Croton L.—croton (native plant)

Euphorbia L.—spurges

    *Euphorbia esula L.—leafy spurge

    **Euphorbia marginata* Pursh—snow-on-the-mountain, ghostweed

    *Euphorbia milii Des Moulins—crown-of-thorns

    *Euphorbia obesa J. D. Hooker—baseball plant, gingham golf ball

    **Euphorbia pulcherrima* Willdenow ex Klotzsch—poinsettia

    **Euphorbia tirucalli* L.—pencil bush, pencil tree, finger tree, milk bush

Hevea Aublet—rubber tree

Jatropha L.—physic nut, coral plant

Manihot Miller—cassava, tapioca

Ricinus L.—castor bean

    *Ricinus communis L.—castor bean

* illustrated

Euphorbiaceae, The Spurge Family

The flowers of Euphorbiaceae are much smaller than the colored bracts that surround them. The flowers are actually inflorescences, groupings of staminate flowers that are reduced to single stamens and pistillate flowers that are solely three-carpellate pistils. The pistils have a stalk (→, arrow) that elongates as the fruit develops. Several family members have milky sap, including the rubber tree. In some, the sap is caustic.

Snow-on-the-mountain (↑), *Euphorbia marginata*

The castor bean (→), *Ricinus communis*, has bright red stigmas and cream-colored stamens. Its seeds contain a deadly poison called ricin.

(↑) The floral bracts of poinsettias, *Euphorbia pulcherrima*, are often mistaken for petals. The real flowers are in the cup-like inflorescences called cyathia. The cups have yellow nectar glands on the rim. The stigmas of this variety are red and the anthers are yellow.

Euphorbias have a wide variety of forms. Several Old World species are adapted to deserts and resemble cacti.

Crown of thorns (→), *Euphorbia milii*

Baseball plant (↘), *E. obesa*

Pencilbush (↓), *E. tirucalli*

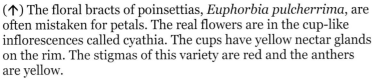

Leafy spurge (↓), *Euphorbia esula*, is an aggressive weed that displaces native plant communities.

Eudicots

Eurosids I

Malpighiales Martius

Linaceae Perleb   12 genera/300 species

Genera include:

Adenolinum Reichbach—flax

> **Adenolinum grandiflorum* (Desfontaines) W. A. Weber—scarlet flax, flowering flax, red flax

> **Adenolinum lewisii* (Pursh) A. & D. Löve—prairie flax, wild blue flax

Hesperolinon (A. Gray) Small—dwarf flax

Linum L.—flax

> **Linum flavum* L.—golden flax, yellow flax

> *Linum usitatissimum* L.—cultivated flax

Mesynium Rafinesque—flax

> **Mesynium rigidum* (Pursh) A. & D. Löve—orange bowls, stiffstem flax

Reinwardtia Dumortier—yellow flax

* illustrated

Linaceae, The Flax Family

Flowers of Linaceae have five petals, which fall off soon after the flower blooms. They also have five sepals, which remain and cover the developing fruit. There are five stamens and usually a divided style. The flowers are borne on long flowering shoots that contain strong fibers. People have used the fibers of *Linum usitatissimum* for at least 10,000 years to make linen. The cloth's name, like the genus name, comes from the Latin word for flax, *linum*. Flax seed and its high omega-3 oil are used for food. It is also the source of linseed oil, which is used in the manufacture of paints and linoleum.

Prairie flax (↑), *Adenolinum lewisii,* in bloom and its dried, split capsules (→). It is used in restoration of roadsides.

Scarlet flax (←), *Adenolinum grandiflorum,* is an annual that is often included in wildflower seed mixes.

(↓) Orange bowls, *Mesynium rigidum,* is a wildflower of the western United States.

Golden flax, *Linum flavum* (↓), is often called yellow flax, but that name is also used for fall-blooming yellow flax of genus *Reinwardtia.*

Eudicots

Eurosids I

Malpighiales Martius

Passifloraceae Roussel

17 genera/670 species

Genera include:

Passiflora L.—passionflower

*Passiflora incarnata L.—purple passionflower

*Passiflora membranacea Bentham

*Passiflora vitifolia Kunth—crimson passionflower, grape-leaf passion-flower, perfumed passionflower

* illustrated

Perfumed or crimson passionflower (←), *Passiflora vitifolia*, has green anthers and pink styles that end in fat white stigmas.

Passifloraceae is a family of vines with very distinctive flowers. They usually have one or more rows of fringe-like corona above the petals. The stamens and pistils are attached to a raised stalk in the middle of the flower. The filaments of the five stamens attach to the middle of the broad anthers. At the end of the stalk, three styles and stigmas project from the ovary. The fruit is a berry. The leaves have a variety of forms, but many are three-lobed.

Passionflowers usually have five sepals and five petals. The sepals are colored like the petals and may be fused at their bases.

Passiflora membranacea (↓) is native to southern Mexico and Central America. As it illustrates, passionflowers come in a variety of forms. It has large purple-red bracts at the base of the flower. The petals and sepals are cream-colored. The leaves are oval, without lobes.

Purple passionflower or maypops (↑), *Passiflora incarnata*, is native from Texas to the east coast and north to southern Pennsylvania.

Fruits of passionflowers have a tough outer covering that surrounds their many flattened seeds. The seeds have a delicious, jelly-like covering, which is the only edible part of the fruit.

Immature fruit of *P. incarnata* (→)

163

Eudicots

Eurosids I

Malpighiales Martius

    Salicaceae Mirbel   55 genera/1010 species

        Genera include:

        Populus L.—cottonwood, poplar

                *Populus angustifolia E. James—narrowleaf cottonwood

                *Populus deltoides Bartram ex Marshall—eastern cottonwood, alamo

        *Salix L.—willow

* illustrated

Salicaceae, The Willow Family

The Salicaceae are a family of dioecious trees and shrubs. The flowers are borne in catkins. Despite their lack of perianth, the flowers of genus *Salix*, the willows, are usually insect-pollinated. They usually have bright yellow anthers. In contrast, the flowers of *Populus* trees are wind-pollinated. Their anthers are usually reddish. The pistillate flowers produce tiny seeds with abundant fibers that carry the seeds away on the wind. *Populus* trees also reproduce vegetatively, by suckering from shallow, horizontal roots. Aspens are particularly noted for this ability to clone themselves.

Pistillate willow catkins, just blooming (←), with developing fruits (↑), and with maturing fruits (↓) and fuzzy seeds

Staminate willow catkins, just beginning to bloom (→), and with stamens fully extended (↘)

Willows (↑), *Salix* species, have a single bud scale covering their flower buds. Willows are difficult to identify to species. Many hybridize in nature.

(←) Young staminate catkins of a plains cottonwood, *Populus deltoides*, are reddish until the anthers mature. Its characteristic leaf shape (↓) can be seen in fossil impressions that are about 20 million years old.

A narrowleaf cottonwood (↓), *P. angustifolia*, produces abundant fiber from its maturing fruits.

Eudicots

Eurosids I

Malpighiales Martius

Violaceae Batsch   20 genera/800 species

Genera include:

Hybanthus Jacquin—green violet

Viola L.

*Viola adunca Smith—mountain purple violet

*Viola canadensis L.—Canadian white violet

*Viola corsica Nyman—Corsican violet

*Viola nuttallii Pursh—yellow montane violet

*Viola rydbergii Greene—white violet, Rydberg's violet

*Viola tricolor L.—Johnny-jump-up, heart's-ease

*Viola x *wittrockiana* Gams ex Kappert—garden pansy

* illustrated

Violaceae, The Violet and Pansy Family

Violaceae flowers are bilaterally symmetrical and have five petals. The lower petal typically has a nectar spur. Nectar guides, lines on petals that show pollinating insects to the nectar inside, are often present. There are five short, stout stamens fitted tightly around the pistil. Violets produce flowers that are insect-pollinated in the spring, but later in summer they may produce inconspicuous self-pollinating flowers that are described as cleistogamous, meaning literally "closed mating." Some violets have ant-dispersed seeds. These seeds have an edible, external oil body that rewards the transporting insect.

Native violets have a variety of habitats and colors.

The yellow montane violet (→), *Viola nuttallii*, grows on dry slopes.

Rydberg's violet (←), *Viola rydbergii*, grows in dense stands in moist, shady microclimates.

Violet fruits reflect the three-carpellate pistil. They split at maturity, and the margins of the ovary wall curl in and squeeze the seeds until they shoot away. Green and mature fruits of the Canadian white violet (↓), *Viola canadensis*, show the three-fold structure.

Early blue violet (↑), *Viola adunca*, has a prominent nectar spur (←, arrow). Two of its petals have bristle or whisker-like appendages.

Violet family members are favorite horticultural subjects. Johnny-jump-ups, *Viola tricolor*, are natives of Europe (↓). They are one of the species that were crossed to produce garden pansies (↓), *Viola* x *wittrockiana*. The nectar spur (↓) of Corsican violet (↓), *V. Corsica*, protrudes beyond its sepals.

Eudicots

Eurosids I

Oxalidales Heinze

Oxalidaceae R. Browne   6 genera/770 species

Genera include:

Averrhoa L.—starfruit, carambola

Oxalis L.—oxalis, wood sorrel, Cape shamrock, oca

    *Oxalis oregana Nuttall—redwood oxalis

    *Oxalis pes-caprae L.—Bermuda buttercup

    *Oxalis regnellii Miquel—false shamrock, wood sorrel

    *Oxalis stricta L.—common yellow oxalis

* illustrated

Oxalidaceae, The Oxalis Family

Oxalis is the only genus of Oxalidaceae in North America. Its flowers are radially symmetrical, with five equal petals. There are ten stamens in two whorls, with the outer ones shorter than the inner ones. The pistil is usually five-carpellate, with five separate styles. In many species, the styles vary in height from flower to flower. The fruit is a ribbed or angled capsule. The leaves are usually compound, with three leaflets. *Oxalis* leaflets show sleep movements; they fold down at night and open out after sunlight hits them. The plants grow from tubers or rhizomes. The family is named for the oxalic acid in its tissues, which makes its leaves taste sour. Common names for *Oxalis* include sour grass and wood sorrel.

Oxalis pes-caprae (→), known as Bermuda buttercup, is a noxious weed along the California coast. A look into the base of the flower (↑) shows the two levels of stamens, five taller inner ones with five shorter outer ones. This plant reproduces vegetatively using small, bulb-like structures.

Many *Oxalis* species are weedy. Common yellow oxalis (←), *O. stricta*, is a pest of greenhouses and lawns. Its fruits are ribbed capsules (→).

Oxalis species are often sold as shamrocks. A number of species are used for house-plants, such as *Oxalis regnellii* (↓), which has triangular leaflets.

Redwood oxalis (→), *Oxalis oregana*, grows as an understory plant in redwood and Douglas fir forests. It is also used as an ornamental.

Eudicots

Eurosids II

Brassicales Bromhead

Brassicaceae Burnett or Cruciferae Jussieu   337 genera/3350 species

Genera include:

Alyssum L.—perennial alyssum

Arabis L.—arabis, rockcress

Armoracia P. Gaertner, Meyer & Scherbius—horseradish

Aubrieta Adanson—aubrieta

   **Aubrieta deltoidea* (L.) DC.—common aubrieta

Brassica L.—broccoli, Brussels sprouts, cabbage, canola, cauliflower, kale, kohlrabi, mustard greens, pak-choi, turnip, and many herbaceous weeds

   **Brassica oleracea* L.—broccoli (also cauliflower, cabbage, Brussels sprouts, kale, collards, and kohlrabi)

Capsella Medikus—shepherd's purse

Crambe L.—sea kale

   **Crambe cordifolia* Steven—giant kale

Descurainia Webb & Berthelot—tansy mustard

   **Descurainia sophia* (L.) Webb ex Prantl—flixweed

Draba L.—draba, whitlow-wort

Erysimum L.—wallflower

   *Erysimum x *marshallii* (Henfrey) Bois—western wallflower

Herperis L.—rocket, dame's rocket

Iberis L.—perennial candytuff

Lepidium L.—pepperweed

Lobularia Desvaux—sweet alyssum

Lunaria L.—honesty, money plant

Matthiola R. Brown—stock

Physaria (Nuttall) A. Gray—double bladderpod, bladderpod

Raphanus L.—radish

   *Raphanus sativus L.—garden radish

Sisymbrium L.—tumble mustard, hedge mustard

Thlaspi L.—pennycress, mountain candytuff

   *Thlaspi arvense L.—field pennycress

Wasabia Matsumura—wasabi

* illustrated

Mustard Chemicals

Related plants often share the ability to synthesize a particular class of molecules. Members of the mustard family make mustard oils, which have a strong, unpleasant taste. The plant tissues contain a precursor molecule and enzymes that convert them to the active form when the tissues are damaged. This system gives the plant a defense against herbivores. The mustard oils are released when the plant is eaten. This keeps most insects and many mammals away from mustards. The cabbage butterflies and a few other insects have evolved the ability to tolerate mustard oils and exploit mustard family plants.

Brassicaceae or Cruciferae, The Mustard Family

The typical mustard flower has four petals and six stamens. Four of the stamens are longer than the other two, as seen in giant kale (↑), *Crambe cordifolia*. The flowers are two-carpellate. The tall, slender inflorescence keeps growing as long as conditions allow. Family members often grow in the cool temperatures of spring or fall. The cross shape formed by the petals gave this family its original name, Cruciferae.

This is a large family, with many weedy species that often grow on disturbed soil on roadsides and vacant lots. Pennycress (↑), *Thlaspi arvense*, and flixweed (↑), *Descurainia sophia,* are examples. The inflorescence has buds and young flowers at the top, with older flowers and developing fruit below. The fruit is called a silicle if it is short and wide (↑) or a silique if it is long and thin (↑). These fruits have two outer covers over a central partition or septum. The seeds form on both sides of the septum.

This Siberian wallflower (←), *Erysimum* x *marshallii*, has elongating pistils and young siliques below these.

Broccoli, one of the many *Brassica oleracea* varieties, has yellow (↓) flowers, although the shoots are usually picked and eaten while the flowers are in bud. The whole plant with tight clusters of buds is shown below.

Aubrieta deltoidea (↑) is a favorite rock garden plant. Like most members of this family, it is an early spring bloomer.

If radish plants, *Raphanus sativus*, are left in the ground after the root has grown, they put up a tall shoot and bloom. Their flowers have clawed, white petals (←).

Eudicots

Eurosids II

Malvales Dumortier

Malvaceae Jussieu   243 genera/4225 species

Malvoideae Burnett

Genera include:

Abelmoschus Medikus

     *Abelmoschus esculentus (L.) Moench—okra

Abutilon Miller—Chinese bellflower, flowering maple, Chinese jute

Alcea L.—hollyhock

     *Alcea rosea L.—common hollyhock

Althaea L.—marsh mallow

Alyogyne Alefeld—blue hibiscus

Callirhoe Nuttall—wine cups, poppy mallow

Gossypium L.—cotton

     *Gossypium hirsutum L.—upland cotton

Hibiscus L.—hibiscus, rose mallow, rose of Sharon

     *Hibiscus rosa-sinensis L.—Chinese hibiscus

Lavatera L.—tree mallow, annual mallow

Malva L.—mallow, cheeseweed

Sidalcea A. Gray—checkerbloom, miniature hollyhock

*Sphaeralcea A. Saint-Hilaire—globe mallow

* illustrated

Malvaceae and Its Subfamilies

In traditional classifications, the mallow family, Malvaceae, was closely associated with three other families—Bombacaceae, Sterculiaceae, and Tiliaceae. When botanists studied the DNA from these families, they found them so closely related that all four families were combined into an enlarged Malvaceae. In the Angiosperm Phylogeny Group classification, there are nine subfamilies in the enlarged mallow family. Two of the subfamilies, Malvoideae and Tilioideae, have temperate members. They are included in this tour. All the subfamilies have tropical members. We visit one tropical subfamily, Byttnerioideae, which holds cacao, the source of chocolate, a favorite food of people around the world. See Appendix A for a more complete listing of Malvaceae subfamilies.

There is still a great deal of study to be done to sort out all the genera in Malvaceae, and it is quite possible that further rearrangement of the subgroups will occur when more information is available about its members.

Cultivars of Chinese hibiscus (←), *Hibiscus rosa-sinensis*, are available in many colors. This ornamental shrub is used in semitropical and mild winter climates.

The globe mallows (→), genus *Sphaeralcea*, are common wildflowers of the western United States.

Fused parts characterize the flowers of the mallow subfamily. The stamens have the bases of their filaments fused into a tube that wraps around the pistil. The bases of the five petals are also fused with the filaments. These cover the ovary. The end of the style and the stigmas are the only part of the pistil that is visible (↑, arrow). The style usually has five branches. The stigmas may be knobs or hair-like. The plants typically have slimy, mucilaginous sap.

Hollyhocks (←), *Alcea rosea*, have been bred in many colors, including red, pink, white, and yellow. The stamens mature first, followed by the filamentous pistils (↓, pink structures). The fruits (↓) are schizocarps.

Okra, *Abelmoschus esculentus* (↓), is eaten when the fruits are immature, like the green pod below.

(↖) The flowers of cotton, *Gossypium hirsutum*, open with white or cream petals that redden after they are pollinated. The fruits are commonly called bolls. The green boll (←) still has the leafy epicalyx that surrounded the flower bud. The mature boll is a capsule (↑) that holds four to five sections of white fibers and seeds.

Eudicots

Eurosids II

Malvales

Eudicots

Eurosids II

Malvales Dumortier

Malvaceae Jussieu

Tilioideae Arnott

Genera include:

Tilia L.—lime tree, linden tree, basswood tree

*Tilia americana L.—American linden, basswood, American lime tree

Eudicots

Eurosids II

Malvales Dumortier

Malvaceae Jussieu

Byttnerioideae Arnott

Genera include:

Ayenia L.—ayenia

Theobroma L.

*Theobroma cacao L.—cacao tree

* illustrated

Malvaceae, The Mallow Family: Tilioideae, The Tilia Subfamily

American basswood or linden tree,
Tilia americana

Tilia trees have clusters of blossoms with a large, light green bract above each inflorescence. (↑) In the fall, the bracts turn brown. When the fruits mature, the bract and fruits blow away as a unit.

Tilia flowers are not large and showy, but they attract abundant bees and other pollinators with their sweet fragrance. The filaments of the stamens are not fused, as they are in the mallow subfamily.

Tilia americana is native to the eastern half of the United States. Several other species of *Tilia* have been imported from Europe for use as ornamentals.

Malvaceae, The Mallow Family: Byttnerioideae, The Byttneria Subfamily

The cacao tree, *Theobroma cacao*, is the source of chocolate. Its flowers grow directly on its trunk and main branches, a phenomenon called cauliflory. Tiny midges pollinate the flowers. The fruits are large pods, which turn yellow or orange when they are ripe. The chocolate "beans" are the seeds inside. The pods must be cut from the trees, since they do not fall on their own. To produce chocolate, the seeds and their pulp are removed from the fruit and fermented for several days. This produces the rich chocolate flavor. The seeds are then dried and processed further.

Cacao flowers (←), young fruit (↙), and mature fruit (↓)

Eudicots

Eurosids II

Sapindales Dumortier

Anacardiaceae R. Brown   70 genera/985 species

Genera include:

Anacardium L.—cashew
>   **Anacardium occidentale* L.—cashew nut

Continus Miller—smoke tree, smoke bush

Mangifera L.—mango

Metopium P. Browne—Florida poisontree, poisonwood

Pistacia L.—pistachio tree

Rhus L.—sumac, poison ivy, poison oak, varnish tree
>   **Rhus diversiloba* Torrey & Gray—Pacific poison oak
>   *Rhus radicans* L.—eastern poison ivy
>   **Rhus rydbergii* Small ex Rydberg—western poison ivy
>   **Rhus typhina* L.—staghorn sumac

Schinus L.—peppertree

Toxicodendron Miller is often included in *Rhus* L.

* illustrated

Anacardiaceae, The Sumac Family

The flowers of Anacardiaceae are small and unisexual. The plants are usually dioecious. A number of plants in the sumac family contain substances that cause skin rashes in people. Some people are more sensitive to these substances than others, which explains why some people get skin rashes when they peel mangoes and others do not. The sap of this family is black or turns black when it dries.

Staghorn sumac (↑, →),
Rhus typhina

Pacific poison oak (↑), *Rhus diversiloba*, has compound leaves with three leaflets. The leaf surface has an oily sheen, which warns of its skin irritants. Western poison ivy (↑), *Rhus rydbergii*, has the typical leaflets in three of the poison ivy–poison oak group. Its fruits are white, ribbed, and berry-like (↑). Its eastern counterpart, *Rhus radicans*, has similar foliage, but is more capable of climbing or vining.

(←) Cashew nuts are the seeds of *Anacardium occidentale*. They develop in a kidney-shaped fruit at the end of an enlarged pedicel. When the fruit is mature, the pedicel becomes fleshy and red. Botanically it is known as a pseudofruit. It is commonly called a cashew apple and is eaten fresh. Cashew apples have a very high vitamin C content. The mature cashew nut is actually the seed inside the tan shell. Cashew nuts are always sold without their shells because the oil in the shell can produce blisters on people's skin. The shells are heated to extract the oil, known as cashew nut shell liquid (CNSL), which has a number of industrial uses.

177

Eudicots

Eurosids II

Sapindales Dumortier

Rutaceae Jussieu   161 genera/1815 species

Genera include:

Amyris P. Browne—torchwood

Choisya Kunth—Mexican orange

Citrus L.—orange, lemon, lime, grapefruit, tangerine

　　Citrus x *limon* (L.) Burman f.—lemon

Coleonema Bartling & H. L. Wendland—confetti bush, breath of heaven

Dictamnus L.

　　Dictamnus albus L.—gas plant, burning bush, fraxinella

Fortunella Swingle—kumquat

Phellodendron Ruprecht—corktree

Poncirus Rafinesque—trifoliate orange

Ptelea L.—hoptree

　　Ptelea trifoliata L.—common hoptree, wafer ash

Ruta L.

　　Ruta graveolens L.—garden rue, herb-of-grace

Skimmia Thunberg—skimmia bush

Zanthoxylum L.—prickly-ash, Hercules club, toothache tree

* illustrated

Rutaceae, The Citrus Family

The citrus family is known for its aromatic, ethereal oils. They are what give citrus fruits their fragrance. The fruits of this family vary widely between species, but many have oil glands in their outer layer. The sections of a citrus fruit such as an orange are each derived from one of the carpels of the pistil. The leaves are often trifoliate, which is compound with three leaflets. The flowers usually have four to five sepals and petals.

Gas plant (←), *Dictamnus albus*, was named because the fumes it gives off in still, warm air can be lit with a match. The fruits of gas plant are covered in red oil glands (←, right photo).

A lemon flower (↑), *Citrus* x *limon*, and the young fruits (↑) show features of the genus. Citrus flowers have their stamens arranged in a ring, and there is a disk at the ovary base. The stigma is knob-like with a dimple in the middle.

Rue, or herb-of-grace (→), *Ruta graveolens*, has flowers with parts in multiples of four or five. There are eight to 10 stamens. The ovaries are deeply lobed and have glands dotting their outer surface.

The hoptree or wafer ash (↓), *Ptelea trifoliata*, has small, greenish white flowers (↓, inset) that can have four or five petals. The fruits (↓) are samaras, and look much like elm fruits.

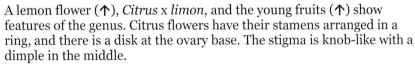
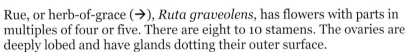

Eudicots

Eurosids II

Sapindales Dumortier

Sapindaceae Jussieu   135 genera/1580 species

Sapindoideae Burnett

Genera include:

Exothea Macfadyen—inkwood

Koelreuteria Laxmann—Chinese flame tree, flamegold tree, goldenrain tree

    **Koelreuteria paniculata* Laxman—goldenrain tree

Litchi Sonnerat—lychee

Sapindus L.—soapberry

    **Sapindus saponaria* L. var. *drummondii* (Hooker & Arnott) L. Benson—
        western soapberry

* illustrated

Soapberry Family Revisions

The traditional classification placed horse chestnuts, soapberry trees, and maples in their own families, but recent studies show that these are all closely related. Maples and horse chestnuts now make up a subfamily, Hippocastanoideae, of the Sapindaceae. A second main group is the soapberry subfamily, Sapindoideae, which is mainly tropical. Its North American representatives are the native western soapberry and a Chinese import, the goldenrain tree. Members of the soapberry subfamily have alternate, pinnately compound leaves. Maples and horse chestnuts, members of the Hippocastanoideae, have opposite leaves.

A Chinese shrub, *Xanthoceras,* makes up another branch of the soapberry family. It is sometimes grown as an ornamental in North America. The last subfamily, Dodonaeoideae, is composed of shrubs that grow in tropical or warm temperate climates and are mostly Australian natives.

Sapindaceae, The Soapberry, Maple, and Horse Chestnut Family: Sapindoideae, The Soapberry Subfamily

Goldenrain tree (↑), *Koelreuteria paniculata,* has bisexual flowers and staminate flowers in one inflorescence (↑). The ovaries develop into three-sided capsules. These papery, inflated capsules (↗, →) break into three segments, each attached to one or two seeds. The light, thin capsule wall aids in wind dispersal of the seed.

This tree is native to China. It was introduced into the United States as an ornamental and has naturalized in many areas. Its abundant seeds allow it to spread. In Florida and similar climates, it is considered to be an invasive pest that threatens native plant communities.

Overview of mature capsules and foliage (→)

Western soapberry (↑), *Sapindus saponaria* var. *drummondii,* has fruits with a translucent outer layer. This is a dioecious species that is native to the south-central and southwestern United States and northern Mexico. Soapberries are named for the saponins in their fruits and other tissues. Saponins are a class of molecules that foam like detergents when they are mixed with water. Sapinaceae is not the only family that makes saponins, but these substances give soapberries their common name.

Eudicots

Eurosids II

Sapindales Dumortier

Sapindaceae Jussieu

Hippocastanoideae Burnett

Genera include:

Acer L.

Acer ginnala Maximowicz—Amur maple, ginnala maple

Acer grandidentatum Nuttall—big tooth maple

Acer negundo L.—box elder

Acer platanoides L.—Norway maple

Aesculus L.

Aesculus glabra Willdenow—Ohio buckeye

* illustrated

Sapindaceae, The Soapberry, Maple, and Horse Chestnut Family: Hippocastanoideae: The Maple and Horse Chestnut Subfamily

Ohio buckeye trees (←), *Aesculus glabra*, have palmately compound leaves. The tall, pyramidal inflorescences arise on the ends of the new growth. The stamens are curved and stick out past the petals. Some of the flowers have only stamens. Others are bisexual. The fruits (→) have a prickly outer layer. Mature fruits (↓) contain a single large seed.

Maples usually have simple leaves with pointed lobes, like the bigtooth maple (←), *Acer grandidentatum*. Maples can also have compound or unlobed leaves, but even then, the margins are toothed. Their fruits are samaras, borne in pairs (→).

Amur maple, *A. ginnala* (→)

Norway maple, *A. platanoides* (↘)

The angle between the paired samaras is a feature that can aid in identifying maples.

Box elder trees (←), *Acer negundo*, have (1) their staminate flowers and (2) pistillate flowers on separate trees. The pistillate flowers have two long, white stigmas. The leaves are compound, but the characteristic samaras, called maple keys, show box elder is a member of *Acer*. The opposite leaf arrangement fits the maple picture, as well.

Tree Diagram of Flowering Plant Orders: Asterids

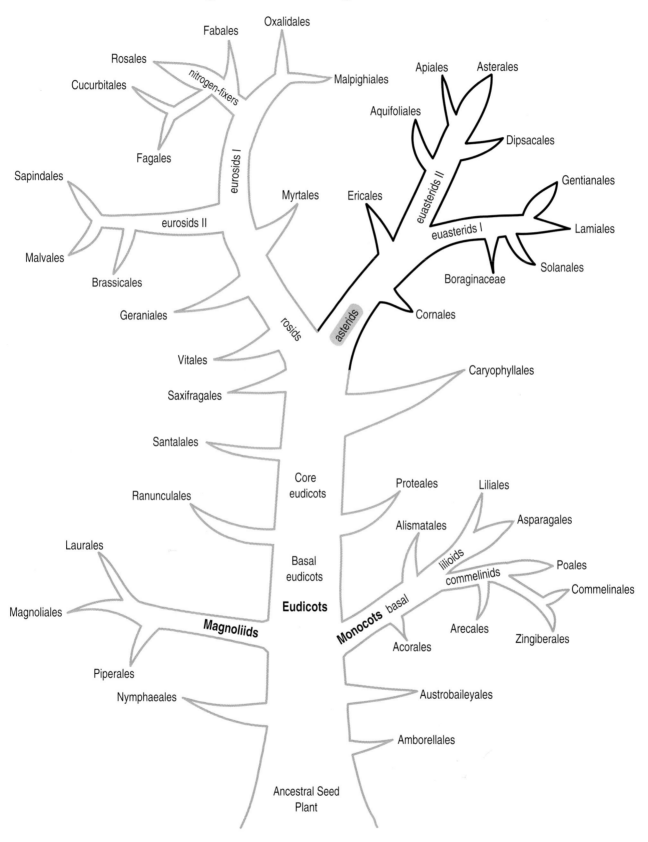

Introducing the Asterids

The asterids are a major branch of the core eudicots, with a third of the angiosperm species. There are four major subgroups—the Cornales order, the Ericales order, euasterids I, and euasterids II.

Cornales includes the hydrangea family, the dogwood family, and the stickleaf family. Ericales is a large order that includes the heath, phlox, primrose, and camellia families. Euasterids I, also called the lamiids, includes Gentianales with the dogbane, gentian, and madder families. Lamiales, another large order of euasterids I, holds the acanthus, mint, olive, and plantain families, and Solanales holds the nightshade and morning glory families. The borage family, Boraginaceae, is not assigned to an order, but is also a part of euasterids I. Euasterids II, also called the campanulids, include the Aquifoliales with the holly family, Apiales with the carrot family, Asterales with the sunflower and bellflower families, and Dipsacales with the honeysuckle and teasel families.

As a group, asterid flowering plants are easier to describe than the other main branch, the rosids. The photo above, a petunia flower with the corolla split and folded back, illustrates some features that euasterids have in common. The petals are fused, which led to an older name for this group, the Sympetalae. The number of stamens equals the number of petals, and the stamens are usually joined to the fused petals. The pistil is usually two-carpellate.

Cornales and Ericales include plants that do not have fused petals and some that have numerous stamens.

The asterids include trees, shrubs, vines, herbaceous plants, and aquatics. Some members of the dogbane family, Apocynaceae, are succulents. Insectivorous plants occur in the Ericales.

Eudicots

Asterids

Cornales Dumortier

Cornaceae Dumortier   2 genera/85 species

Genera include:

Cornus L.—dogwood

    *Cornus canadensis L.—bunchberry

    *Cornus florida L.—flowering dogwood

    *Cornus kousa Hance—Kousa dogwood

    *Cornus sericea L.—redtwig or red-osier dogwood

* illustrated

Cornaceae, The Dogwood Family

(←) *Cornus florida* is native to the eastern United States. It and other flowering dogwoods have flower clusters that are surrounded by four large bracts that may be white, cream, or pink.

The leaves of Cornaceae are usually simple with entire margins. They are usually opposite, although bunchberries have whorled leaves. The secondary veins of the leaves arch toward the margins (↑). The flowers have four petals and four stamens. They form in clusters and open nearly simultaneously. The fruits are small drupes and are white, red, blue, or blue-black in various species. In some species, the fruits of a flower cluster fuse together and form a multiple fruit.

Two cultivars of Kousa dogwood (↑, ↗), *Cornus kousa*, have bracts with slightly different colors and shapes. The ball of flowers at the center of the bracts (↑) develops into a cluster of fruits.

Bunchberry, *Cornus canadensis*, is a low, groundcover plant that has red fruits (→). It lives in moist coniferous forests.

The redtwig or red-osier dogwood, *Cornus sericea*, has no bracts around its flowers (↓). The fruits are white or blue (↓) and berry-like.

Eudicots

Asterids

Cornales Dumortier

Hydrangeaceae Dumortier   17 genera/190 species

Genera include:

Carpenteria Torrey—tree anemone

Decumaria L.—climbing hydrangea

Deutzia Thunberg—deutzia

Fendlera Engelmann & A. Gray—fendlerbush

Hydrangea L.—hydrangea

      Hydrangea macrophylla (Thunberg) Seringe—garden hydrangea

Jamesia Torrey & A. Gray

      Jamesia americana Torrey & A. Gray—waxflower, cliffbush

Philadelphus L.—mock orange, Indian arrowwood

Schizophragma Seibold & Zuccarini—Japanese hydrangea vine

* illustrated

Hydrangeaceae, The Hydrangea Family

Mock orange, *Philadelphus*, is named for the sweet, orange-like fragrance of its flowers. The ovary begins to develop before the petals have dropped (↗). The fruits are backed by the persistent sepals (↑).

The leaves of Hydrangeaceae are opposite and joined across the node by a line. The flowers have at least twice as many stamens as petals. There are two to five styles and stigmas. The flowers are bisexual in most genera. The hydrangeas have sterile flowers that attract pollinators. These surround the small, fertile, seed-producing flowers in native hydrangeas. In hybrid garden hydrangeas, like the one below, almost all the flowers are sterile. In many cultivars, the flower color reflects the pH of the soil. In alkaline soil, the flowers are blue; in acid soil, they are pink.

There are many cultivars of garden hydrangeas (↑), *Hydrangea macrophylla*. The lacecap cultivars have tiny fertile flowers in the center (↑) and large, showy sterile flowers around the outside. The mophead cultivars have only the larger, sterile flowers (↑).

Waxflower (←), *Jamesia americana*, is a shrub that is native to the southwestern United States.

Eudicots

Asterids

Cornales Dumortier

Loasaceae Jussieu

14 genera/265 species

Genera include:

Eucnide Zuccarini—rock nettle, stingbush

Mentzelia L.—blazingstar, stickleaf

> *Mentzelia albicaulis* Douglas ex Hooker—white-stemmed stickleaf
>
> *Mentzelia nuda* (Pursh) Torrey & A. Gray—prairie evening star, bractless blazingstar, sand lily
>
> *Mentzelia reverchonii* (Urban & Gilg) H. J. Thompson & Zavortink—yellow stickleaf

Petalonyx A. Gray—sandpaper plant

* illustrated

Loasaceae, The Blazingstar or Stickleaf Family

The flower (←) and fruits (→) of yellow stickleaf, *Mentzelia reverchonii*

The flowers of the blazingstar family are usually large and showy. They have many stamens of various lengths. In some species, there are enlarged, flattened stamens that look like petals. There is a single pistil with one style and an inferior ovary. The fruit is a roughly cylindrical capsule that has the persistent sepals around the top rim. These plants are well adapted to arid or semi-arid areas.

The leaves of Loasaceae have barbed hairs that often contain silica. This makes them feel like sandpaper. Some species have stinging hairs, so use caution if you wish to feel the leaves.

White-stemmed stickleaf (←), *Mentzelia albicaulis*, has much smaller flowers than its relatives. Even though its flowers aren't typical, its fruits and the sandpaper feel of its leaves identify it as a member of *Mentzelia*.

Like many blazingstars, the flowers of the bractless blazingstar (↓), *Mentzelia nuda*, open in the late afternoon and stay open all night, an adaptation for hot, arid climates. The fruit retains the single style in the center and the five sepals around the top rim (→).

191

Eudicots

Asterids

Ericales Dumortier

Balsaminaceae Berchtold & J. Presl   2 genera/1000 species

Genera include:

Impatiens L.—busy lizzie, touch-me-not, balsam, impatiens, jewel weed

Impatiens balfourii Hooker f.—Balfour's impatiens, poor man's orchid

Impatiens niamniamensis Gilg—Congo cockatoo, African King, African Queen

Impatiens walleriana Hooker f.—busy lizzie, sultana, bedding impatiens

* illustrated

Balsaminaceae, The Impatiens Family

Busy lizzie or bedding impatiens (↑), *Impatiens walleriana*, is grown as a summer annual. It is well adapted for shady, understory environments. This Balfour's impatiens (↑), *Impatiens balfourii*, shows the nectar spur on the bud (arrow) and has developing green fruits.

Impatiens is the only genus of Balsaminaceae in North America. There is only one other in the family, a southeast Asian genus with one species. *Impatiens* includes a number of imported garden flowers and about six native species. The annuals have fleshy, translucent stems. Their leaves often have toothed margins. The flowers have three sepals, one of which forms a nectar spur, a long, curved, hollow tube that holds nectar. The flowers are bilaterally symmetrical, with a corolla of five petals. In some species, a pair of joined petals on either side makes the flower appear to have less than five petals. The anthers are fused and form a cap over the stigma. The fruit is a capsule. In some species, commonly known as touch-me-nots, this capsule literally pops when it is touched, quickly splitting into curled segments and flinging its seeds away.

The sepal that forms the nectar spur is much larger than the other two (→). Only one of the small sepals is visible in this photograph.

Congo cockatoo or parrot plant (←), *Impatiens niamniamensis*, has lime green petals that are fused to the yellow and scarlet nectar spur.

Eudicots

Asterids

Ericales Dumortier

Ericaceae Jussieu   126 genera/3995 species

Genera include:

Andromeda L.—bog rosemary

Arbutus L.—madrone, strawberry tree

Arctostaphylos Adanson—kinnikinnick, manzanita, bearberry

　　　Arctostaphylos uva-ursi (L.) Srengel—kinnikinnic, bearberry

Calluna Salisbury—heather

Cassiope D. Don—mountain heather, moss heather

Chamaedaphne Moench—leatherleaf

Chimaphila Pursh—pipsissewa, prince's pine

Empetrum L.—crowberry

Epigaea L.—trailing arbutus

Erica L.—heath

　　　Erica carnea L.—winter heath, alpine heath

Gaultheria L.—wintergreen, salal, snowberry, tea berry

　　　Gaultheria shallon Pursh—salal

Kalmia L.—mountain laurel, alpine laurel, bog laurel, sheep laurel, lambkill

　　　Kalmia latifolia L.—mountain laurel, calico bush

Moneses Gray—one-flowered wintergreen

Monotropa L.—Indian pipe, pinesap

Phyllodoce Salisbury—mountain heath

Pieris D. Don—fetterbush, lily-of-the-valley bush

Pterospora Nuttall

　　　Pterospora andromedea Nuttall—pinedrops

Pyrola L.—wintergreen, shinleaf

Rhododendron L.—rhododendron, azalea, Labrador tea

Vaccinium L.—blueberry, cranberry, huckleberry, lingonberry, whortleberry

　　　Vaccinium corymbosum L.—highbush blueberry

* illustrated

Ericaceae, The Heath and Heather Family

Winter heath (↑), *Erica carnea*   Kinnikinnick flowers (↑), *Arctostaphylos uva-ursi*, and fruits (↑)

Salal flowers and fruits (↓), *Gaultheria shallon*   Rhododendron, *Rhododendron* hybrid (↓)

Highbush blueberry fruits (↓), *Vaccinium corymbosum*

Ericaceae is a large family that is characteristic of moist, acidic forest soils, yet has many members that live in other habitats. They are common components of tundra biota, and live on dry, sunny slopes in the western montane life zone. The acid bog dwellers have special mycorrhizae that enable them to survive in that harsh environment. There are a variety of flower shapes in the family. Urn-shaped flowers are unique to it. Other shapes include tubular, funnel-like, open bowls, little bells with tiny rounded lobes, and even separate petals in a few. The leaves of many species are smooth, thick, leathery, and evergreen. Some, such as heather and heath, have narrow, needle-like leaves (upper left photo).

Pinedrops (←), *Pterospora andromedea*, is a parasite that has no chlorophyll. It lives with the aid of a fungus that connects its roots to a host plant. Another member of the parasite group, Indian pipes, genus *Monotropa*, is pure white.

Mountain laurel (→), *Kalmia latifolia*, has bowl-shaped flowers. The 10 stamens are tucked into pocket on the corolla until a pollinator lands (upper flower). Then they pop up (lower flower) and deposit their pollen on the insect. Kalmias are poisonous to domestic animals and humans, but deer can eat them without being harmed.

Eudicots

Asterids

Ericales Dumortier

Polemoniaceae Jussieu   18 genera/385 species

Genera include:

Cobaea Cavanilles—cup-and-saucer vine

Collomia Nuttall—collomia, tiny trumpet

Gilia Ruiz & Pavón—gilia, blue bowls, bird's-eye gilia, globe gilia, thimble flower

Ipomopsis Michaux—ipomopsis, skyrocket, trumpet gilia

   *Ipomopsis aggregata (Pursh) V. Grant—scarlet gilia, fairy trumpet

Linanthus Bentham—linanthus, babystars, whisker-brush

Phlox L.

   *Phlox condensata (A. Gray) E. Nelson—alpine phlox, cushion phlox

   *Phlox paniculata L.—summer phlox, perennial phlox

   *Phlox subulata L.—moss pink, creeping phlox

Polemonium L.—polemonium, Jacob's ladder

   *Polemonium caeruleum L.—Jacob's ladder, charity

   *Polemonium viscosum Nuttall—sky pilot

* illustrated

Polemoniaceae, The Phlox Family

Garden species of the phlox family:
Summer phlox (↑), *Phlox paniculata*

Moss pink (↑), *Phlox subulata*

Jacob's ladder (→), *Polemonium caeruleum*

Alpine phlox (↑), *Phlox condensata*, and sky pilot (↑), *Polemonium viscosum*, grow in the alpine tundra environment.

Scarlet gilia, also called fairy trumpets, *Ipomopsis aggregata*, has populations with brilliant red flowers (↓). Other populations have white flowers. Where the ranges of the two overlap, hybrids occur in several shades of pink (→).

The flowers of Polemoniaceae have fused sepals and petals. In most, the corolla is narrow at the base and flares out at the top, ending in five points or lobes. Some genera, like *Polemonium*, have bowl-shaped flowers. The stamens of some species are different lengths or are attached to the corolla at different heights, which encourages cross-pollination. The style elongates after the pollen is shed. It has three branches on which the stigmas form. The flowers are often in clusters. The polemoniums are known for their skunky odor. Many polemoniums and phlox species grow at high altitudes or latitudes.

Eudicots

Asterids

Ericales Dumortier

Primulaceae Borkhausen   9 genera/900 species

Genera include:

Androsace L.—rock jasmine

Dodecatheon L.—shootingstar

    **Dodecatheon pulchellum* (Rafinesque) Merrill—shootingstar

Hottonia L.—featherfoil, water violet

Primula L.—primrose

    **Primula denticulata* Smith—drumstick primrose

    *Primula parryi A. Gray—Parry's primrose

    *Primula x *polyantha* Miller—polyanthus primrose

    *Primula veris L.—cowslip

    *Primula vialii Delavay ex Franchet—orchid primrose, poker primrose

* illustrated

Primulaceae, The Primrose Family

Shootingstar (←), *Dodecatheon pulchellum*, is native to the western United States.

Polyanthus primroses (→), *Primula* x *polyantha*, are ornamentals that are available in many colors.

The plants of Primulaceae have a basal rosette of leaves. The flowers form in umbels on a tall, leafless stem called a scape. They have fused sepals. The petals are fused into a tube at their bases. Their five lobes open flat or bend back toward the stem. In many primroses, there are stamens and styles of different heights. Some flowers have short stamens and a long style, and are known as pin flowers. Others of the same species have long stamens and a short style, and are known as thrum flowers. These flower structures promote cross-pollination.

Parry's primrose (←), *Primula parryi*, grows near streams in the Rocky Mountains. It is very showy, with individual flowers about three centimeters across.

The cowslip (→), *Primula veris*, has a long history of cultivation in Europe, where it originated. It has naturalized in the north-eastern United States

The drumstick primrose (↓), *Primula denticulata*, is also called the Himalayan primrose.

Primula vialii (→), sometimes called the orchid primrose, is a native of China. Its red calyxes provide a bright contrast to the pale purple flowers. The tall, slender inflorescence is unusual in the primrose family.

Eudicots

Asterids

Ericales Dumortier

Sarraceniaceae Dumortier   3 genera/15 species

Genera include:

Darlingtonia Torrey

   **Darlingtonia california* Torrey—California pitcher plant, cobra lily

Sarracenia L.—pitcher plant

   **Sarracenia flava* L.—yellow pitcher plant

   **Sarracenia purpurea* L.—purple pitcher plant

* illustrated

A Tale of Three Pitcher Plants

As remarkable as *Darlingtonia* and *Sarracenia* are, they are not the only pitcher plants. Two other families of pitcher plants exist. The Asian pitcher plants are *Nepenthes*, the sole genus of family Nepenthaceae, in the order Caryophyllales. The oxalis order, Oxalidales, has the family of Australian pitcher plants, Cephalotaceae, with one genus, *Cephalotus*. These three families are obviously not closely related, since they fall on three separate branches of the core eudicots, yet all have leaves adapted for trapping insects and absorbing nutrients from them.

This is an example of convergent evolution, the similar appearance in organisms that adapt to similar conditions. The three pitcher plants developed similar solutions to the same environmental stress. They grow in boggy soils where nitrogen and other minerals are limiting nutrients. Any plant that can procure these minerals has a survival advantage.

Each of the pitcher plants has unique features. The pitchers look very similar superficially, but they have differences in structure. In Nepenthaceae, pitchers hang from a tendril on the end of a normal-looking leaf. *Cephalotus* has both normal leaves and little pouch-like pitchers. Members of Sarraceniaceae have only pitchers; they do not have any normal-looking leaves. Their pitchers grow in rosettes.

If we look beyond pitcher plants to all insectivorous plants, we see more of the picture. There are four families of insectivorous plants in Caryophyllales, all of which are related. They use several mechanisms to trap insects—snap traps with trigger hairs, sticky flypaper hairs, and pitcher pitfall traps. Their DNA shows that they shared a common ancestor, and it is highly probable that the ancestor developed insectivory. In Ericales, two related families are insectivorous. There is evidence that the order Lamiales, in euasterids I, has five related families that trap insects. *Cephalotus* is the sole insectivorous member of Oxalidales, and the only eurosid that has this ability. From this, it appears that insectivory has evolved at least four times.

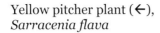

Yellow pitcher plant (←),
Sarracenia flava

Purple pitcher plant (→),
Sarracenia purpurea

The pitcher plant family has highly modified leaves that enable it to live in the nutrient-poor soils of marshes and bogs. The leaves are adapted to trap insects and to absorb nutrients from them, which provides the plant with nitrogen-containing compounds and other minerals.

The pitcher plant flower (←) looks nearly as strange as its leaves. The style has five large, flat branches. The petals drape around this enlarged style. The style and the sepals remain on the developing ovary (↙), looking like five-part umbrellas over it and behind it. Pitcher plant leaves produce nectar that attracts insects. Inside the pitcher, the leaf secretes a slippery wax and has downward pointing hairs. Insects enter the pitcher and fall into a watery liquid at the bottom, where their bodies decompose.

The California pitcher plant or cobra lily, *Darlingtonia california* (↓), attracts insects with nectar. They crawl up the "fang" leaf to an entrance on the bottom curve of the pitcher. When an insect reaches the opening, it sees the light coming through transparent spots in the top of the rounded "head" (↘). It flies up and is trapped, then falls into the pitcher, where bacteria digest it.

Eudicots

Asterids

Ericales

201

Eudicots

Euasterids I

(not placed in an order)

Boraginaceae Jussieu   130 genera/2635 species

Genera include:

Amsinckia Lehm—fiddleneck

Anchusa L.—anchusa, cape forget-me-not

Borago L.

> **Borago officinalis* L.—borage

Cryptantha G. Don—cryptantha, hidden flower, miner's candle

Cynoglossum L.—hound's tongue, Chinese forget-me-not, wild comfrey

> **Cynoglossum officinale* L.—hound's tongue

Echium L.—viper's bugloss, pride of Madeira

Eritrichium Schrader ex Gaudin—alpine forget-me-not

Hackelia Opiz—stickseed, beggar's lice, false forget-me-not

> **Hackelia floribunda* (Lehmann) Johnston—stickseed

Heliotropium L.—heliotrope

Hydrophyllum L.—waterleaf, Shawnee salad

Lappula Moench—stickseed

Lithospermum L.—puccoon, stoneseed

> **Lithospermum multiflorum* Torrey ex A. Gray—puccoon, wayside gromwell

Mertensia Roth—mertensia bluebells, tall chiming bells, Virginia bluebells

> **Mertensia lanceolata* (Pursh) DC.—prairie bluebells

Myosotis L.—forget-me-not

> **Myosotis sylvatica* Ehrhart ex Hoffmann—woodland forget-me-not

Nama L.—fiddleleaf, purple mat, nama

Nemophila Nuttall—baby blue eyes, five-spot, nemophila

*Phacelia Jussieu—phacelia, wild heliotrope, scorpion weed

Pulmonaria L.—lungwort

Symphytum L.—comfrey

NOTE: This family includes the former Hydrophyllaceae, the waterleaf family, which is no longer recognized as a separate family.

* illustrated

Boraginaceae, The Borage and Waterleaf Family

(←) Prairie bluebells, *Mertensia lanceolata*

Flowers of hound's tongue (↑), *Cynoglossum officinale* and its fruits (→), which are four nutlets with hooked bristles

(↓) Puccoon, *Lithospermum multiflorum*

Boraginaceae is not yet assigned to an order within the euasterids I. The flowers have five petals that are fused into a tubular or wheel-like corolla. There are often ridges or bumps near the throat of the corolla tube. Many species have their flowers in coiled inflorescences, with the flowers on the outside of the coil. In some species, the five stamens stick out well beyond the corolla. The pistil is two-carpellate, but develops four locules, each with one ovule. The ovaries mature into four closed, dry fruits called nutlets. (See hound's tongue above.) In forget-me-nots and several others, the flowers change color after they are pollinated.

This *Phacelia* species (←) of the waterleaf subfamily shows the typical coiled inflorescence, which is described botanically as scorpioid or helicoid.

Stickseed (→), *Hackelia floribunda*, has nutlets with little hooked prickles. These cling readily to animals and clothing and aid in seed dispersal.

Borage (←), *Borago officinalis*, is the herb for which the family is named. It has coarse, bristly hairs that are easily seen. These are common in the family.

Forget-me-nots (→), *Myosotis sylvatica*, have a characteristic ring of yellow at their throat that turns to white in older flowers.

Eudicots

Euasterids I

Gentianales Lindley

Apocynaceae Jussieu   415 genera/4555 species

Genera include:

Amsonia Walter—bluestar

Apocynum L.—dogbane, Indian hemp

   **Apocynum androsaemifolium* L.—spreading dogbane

Asclepias L.—milkweed

   **Asclepias asperula* (Decaisne) Woodson—antelope horns, spider milkweed

   **Asclepias speciosa* Torrey—showy milkweed

   **Asclepias tuberosa* L.—butterfly weed

Catharanthus G. Don—Madagascar periwinkle

Hoya R. Brown—waxflower, waxplant, honey flower

   *Hoya carnosa (L. f.) R. Brown—waxflower

Matelea Aublet—milkvine

Nerium L.

   *Nerium oleander L.—oleander

Plumeria L.—frangipangi, plumeria

Stapelia L.—carrion flower, starfish flower

Vinca L.—periwinkle

   *Vinca minor L.—common periwinkle, dwarf periwinkle

NOTE: This family contains the former Asclepiadaceae, the milkweed family, which is no longer recognized as a separate family.

* illustrated

Apocynaceae, The Dogbane and Milkweed Family

Oleander (←), *Nerium oleander*, is used as an ornamental, but carries a potent poison in all its parts.

Waxflower (→), *Hoya carnosa*, looks as if its fleshy flowers are made of plastic. It is grown as a houseplant.

The flowers of this family have a fused, tubular corolla that flares into five lobes. There are fringes at the throat of some species, (↑) as in oleander flowers. Milkweed flowers have hoods, prominent structures above the corolla. The stamens of this family are hard to see because they stick together or to the side of the stigma. The stigma can have a flattened top and be five-sided or two-lobed. The pistil usually has two separate ovaries that share a single stigma. Fruit types include drupes, berries, and follicles. There are often (↓) paired follicles, whose seeds have a tuft of hairs that aids in wind dispersal.

Fruits of spreading dogbane (←), *Apocynum androsaemifolium*, develop from the two separate ovaries of the flower. The shared stigma still joins the ends of these fruits.

The stamens of common periwinkle (→), *Vinca minor*, form a cover over the stigma head. Note the low crown at the throat of the corolla.

Milkweeds are named for their milky sap. Their pollen forms in little packets called pollinia that stick to pollinating insects. The petals often fold back; the five hoods form a star shape above them. The central knob holds the stamens and pistil.

Only a few of the flowers in the inflorescence of butterfly weed (→), *Asclepias tuberosa*, will go on to produce the fruit, a tall, tapered follicle (↘). The follicles each contain hundreds of seeds, each with a sail of silky hairs.

Antelope horns, *Asclepias asperula*, has greenish white petals (↓) with white and reddish purple hoods. Each inflorescence commonly produces two, upturned fruits, hence the plant's common name.

(↓) Showy milkweed, *Asclepias speciosa*, has fragrant flowers with dark pink petals and light hoods. Its large, oval leaves have smooth margins.

Eudicots

Euasterids I

Gentianales Lindley

Gentianaceae Jussieu   87 genera/1655 species

Genera include:

Centaurium Hill—centary, rosita

Eustoma Salisbury ex G. Don—lisianthus

*Eustoma grandiflorum (Rafinesque) Shinners—prairie gentian, tulip gentian

Exacum L.—Persian violet, German violet

Frasera Walter—green gentian, frasera, elk weed

*Frasera speciosa Douglas—green gentian, monument plant

Gentiana L.—gentian

*Gentiana dahurica Fischer—common gentian

Gentianella Moench—dwarf gentian

*Gentianella acuta (Michaux) Hiitonen—little gentian

Gentianopsis Ma—fringed gentian

*Gentianopsis thermalis (Kuntze) Iltis—Rocky Mountain fringed gentian

Obolaria L.—Virginia pennywort

Sabatia Adanson—rose gentian

Swertia L.—star gentian, felwort

* illustrated

Gentianaceae, The Gentian Family

The flowers of the gentian family have four to five petals that are fused, either just at the base or fused into a cup or bell shape. The stamens are attached to the petals, often near the base. The stems often have four ridges, called wings, that run their length. The pistil is two-carpellate, and has a superior ovary. There is one style, with one stigma that may have two lobes.

Prairie gentian or tulip gentian (←), *Eustoma grandiflorum*, is a native to grasslands of the midwestern United States. Horticultural varieties have been bred, including double-flowered ones with extra petals (→).

(←) When this flower blooms, the stamens mature first (lower flower). The two-lobed stigma opens later (upper flower).

Green gentian or monument plant (→), *Frasera speciosa*, has been described as a biennial, but probably spends several years as a rosette plant before it blooms and dies. Its flowers have four greenish white petals with prominent nectar glands.

Little gentian, *Gentianella acuta*, has a fringe of hairs inside the corolla (→).

Fringed gentian, *Gentianopsis thermalis*, is an annual that blooms (↓) in late summer and early fall.

Gentiana dahurica (↓) shows a common feature in this family, small pointed pleats between each of the petal lobes.

Eudicots

Euasterids I

Gentianales Lindley

Rubiaceae Jussieu   600 genera/10,000 species

Genera include:

Bouvardia Salisbury—bouvardia, firecracker bush
Cephalanthus L.—buttonbush, button willow
Cinchona L.—quinine
Coffea L.
> *Coffea arabica* L.—coffee
Galium L.—bedstraw
> *Galium aparine* L.—cleavers, goose grass
> *Galium boreale* L.—northern bedstraw
> *Galium odoratum* (L.) Scopoli—sweet woodruff
Gardenia Ellis—gardenia
Hedyotis L.—star violet, diamond flower
Houstonia L.—bluet, Quaker ladies, Venus' pride
Ixora L.—ixora
> *Ixora coccinea* L.—flame-of-the-woods, jungle flame
Mitchella L.—partridge berry
Pentas Bentham
> *Pentas lanceolata* (Forsskaol) Deflers—pentas, Egyptian star cluster
Psychotria L.—ipecac
Rubia L.—madder

* illustrated

Rubiaceae, The Madder Family

Rubiaceae, which includes coffee, ipecac, and quinine, is comparable in size to the grass family, but most species of this large family are tropical. The leaves of this family are opposite or occasionally whorled. In the whorled species, the extra "leaves" at each node are actually stipules. Only the true leaves have axillary buds at their bases. The flowers have four to five fused petals, which form a slender tube that flares into lobes. The stamens attach to the corolla near the throat of the tube and alternate with the petal lobes. The pistil has a small, inferior ovary. The styles may have different lengths in flowers of the same species, which promotes cross-fertilization.

The genus *Galium* includes native species throughout North America. Square stems and fruits in pairs are distinguishing features. Shown here are:

Northern bedstraw (↖), *G. boreale*

Cleavers or goose grass (←), *G. aparine*, is covered with hooked hairs, including on its fruits.

Sweet woodruff (↓), *G. odoratum*, is a European native. It is a useful garden perennial for shady areas.

Flame-of-the-woods (↑), *Ixora coccinea*, and (↑) star flower, *Pentas lanceolata*, are landscape shrubs that were imported from the Old World tropics.

Coffee "beans" are actually the seeds of this coffee tree (→), *Coffea arabica*. The fruits are drupes that hold two seeds each. Several species of *Coffea* are cultivated to make coffee blends. The plants must be grown in frost-free areas.

Eudicots

Euasterids I

Lamiales Bromhead

Acanthaceae Jussieu   229 genera/3500 species

Genera include:

Acanthus L.—acanthus, bear's breech, oyster plant

    **Acanthus hungaricus* L.—bear's breech

    *Acanthus mollis L.—bear's breech

Aphelandra R. Browne—zebra plant

Avicennia L.—black mangrove

Fittonia Coemans

    **Fittonia verschaffeltii* (Lemaire) van Houtte—nerve plant, mosaic plant

Hypoestes Solander ex R. Browne

    **Hypoestes phyllostachya* Baker—polka dot plant

Justicia L.—shrimp plant

    **Justica brandegeana* Wasshausen & L. B. Smith—red shrimp plant

Ruellia L.—ruellia, Mexican petunia, wild petunia

    *Ruellia humilis Nuttall—fringeleaf wild petunia

Thunbergia Retzius—clock vine, sky flower

    *Thunbergia alata Bojer ex Sims—black-eyed susan vine

* illustrated

Acanthaceae, The Acanthus Family

Ornamentals of the acanthus family include:

Ruellia or wild petunia (↑), *Ruellia humilis*   (↑) Black-eyed susan vine, *Thunbergia alata*

Nerve plant or mosaic plant, *Fittonia verschaffeltii* (↑)

The acanthus family has bracts, either beneath single flowers or in spike-like arrangements with the flowers peeking out from between them. The bracts may be either green or brightly colored. The flowers have corollas of four to five fused petals. Many, like *Justica*, have two-lipped corollas. There are usually four stamens, but some species, like the *Ruellia* above, have only two. There is a two-carpellate pistil with a superior ovary. The fruits of the acanthus family are capsules that explosively release their seeds and fling them away.

Red shrimp plant, (↓) *Justica brandegeana*, has bracts in shades of red. Its flowers are white with red markings visible inside the corolla.

Bear's breech is a common name given to several species of *Acanthus*, including *A. hungaricus* (←) and *A. mollis* (↑).

Polka dot plant (↓), *Hypoestes phyllostachya*, is grown for its beautifully marked foliage.

Eudicots

Euasterids I

Lamiales Bromhead

Bignoniaceae Jussieu   110 genera/800 species

Genera include:

Bignonia L.—crossvine

Campsis Loureiro—trumpet vine, trumpet creeper

  *Campsis radicans (L.) Seemann ex Bureau—trumpet vine, trumpet creeper

Catalpa Scopoli—catalpa

  *Catalpa speciosa (Warder) Warder ex Engelmann—northern catalpa tree

Chilopsis D. Don

  *Chilopsis linearis (Cavanilles) Sweet—desert willow

Incarvillea Jussieu—hardy gloxinia

Jacaranda Jussieu—jacaranda tree

Kigelia DC.—sausage tree

Macfadyena A. DC.†—yellow trumpet vine

Tabebuia Gomes ex DC.—trumpet tree, West Indian boxwood

Tecoma Jussieu—orange bells, Cape honeysuckle

* illustrated
† A. DC. = Alphonse de Candolle

Bignoniaceae, The Trumpet Vine Family

Trumpet vine, *Campsis radicans* (↑), a native of eastern North America, has red-orange flowers. Yellow varieties have been bred for horticultural use (↑). The fruits are long pods with a ridge on either side.

The trumpet vine or trumpet creeper family is characterized by colorful, trumpet-shaped flowers and long, pod-like fruits. Four stamens attach inside the corolla, grouped in two pairs. There may be an additional stamen-like structure, called a staminode, that is nonfunctional. The two-carpellate pistil has a superior ovary. Its stigma typically has two flap-like lobes. These lobes close after contact with a pollinator, which reduces self-pollination. The ovary develops into a fruit that is typically a long, thin pod with two halves. Some family members, like the sausage tree, have short, plump fruits. Inside the pod, the seeds are not at all like beans. Instead, the fruits are packed with flattened seeds that have membranous wings or a fringe of hairs.

Desert willow, *Chilopsis linearis*, is not a true willow. This small tree is native to the southwestern United States. Its flowers (↑) and its fruits (↓) show that it is a member of Bignoniaceae.

Northern catalpa trees (↓), *Catalpa speciosa*, have large white flowers with throat markings. The fruits are very long (↓).

Eudicots

Euasterids I

Lamiales Bromhead

Lamiaceae Martynov or Labiatae Jussieu   236 genera/7173 species

Genera include:

Agastache Clayton ex Gronovius—hummingbird mint, sunset hyssop

Ajuga L.—ajuga, carpet bugleweed

Callicarpa L.—beautyberry

Caryopteris Bunge—bluebeard, blue spirea, blue mist

Clerodendrum L.—clerodendrum, bleeding heart vine, glorybower

Dracocephalum L.—dragonhead

Hyssopus L.—hyssop

Lamium L.—lamium, dead nettle, yellow archangel, henbit

  *Lamium maculatum L.—spotted dead nettle, spotted henbit

Lavandula L.—lavender

Marrubium L.—horehound mint

Melissa L.—lemon balm

Mentha L.—mint, peppermint, spearmint, pennyroyal

Molucella L.—bells of Ireland, shell flower

Monarda L.—horsemint, bergamot, bee balm, monarda

Nepeta L.—catnip, catmint

Ocimum L.—basil

Origanum L.—oregano, marjoram, dittany

Perovskia Karelin—Russian sage

Phlomis L.—Jerusalem sage

  *Phlomis cashmeriana Royle ex Bentham—Kashmir sage

Physostegia Bentham

  *Physostegia virginiana (L.) Bentham—obedient plant, false dragonhead

Plectranthus L'Héritier—Swedish ivy

Prunella L.—self-heal, prunella

Rosmarinus L.—rosemary

Salvia L.—ornamental and culinary sage

  *Salvia darcyi J. Compton—red mountain sage

Satureja L.—summer and winter savory, yerba buena

Scutellaria L.—skullcap mint

  *Scutellaria scordifolia Fischer ex Schrank—skullcap

Solenostemon Thonning—coleus

Stachys L.—betony, stagger-weed

  *Stachys byzantina K. Koch—lamb's ear

Tectona L. f.—teak tree

Teucrium L.—germander

Thymus L.—thyme

  *Thymus praecox Opiz—mother-of-thyme, creeping thyme

Vitex L.—chaste tree

  *Vitex agnus-castus L.—lilac chaste tree, chaste berry

* illustrated

Lamiaceae or Labiatae, The Mint Family

Obedient plant (↑), *Physostegia virginiana*   Lamb's ear (↑), *Stachys byzantina*

Red mountain sage, *Salvia darcyi* (↑)   Dried calyxes of Kashmir sage, *Phlomis cashmeriana* (↑)

The mint family typically has square stems and simple, opposite leaves with serrate margins. Lobed and compound leaves also occur. Its flowers have a tubular corolla that opens in two lips. The flowers are borne in clusters along the stems. The tubular calyx persists after the flower has bloomed. The fruits, usually four nutlets, develop within the calyx.

The ethereal oils of the mint family are one of its most notable characteristics. People have used this family for herbal seasonings since the beginning of recorded history. The oils form in glands on the surface of the leaves. Mint family herbs include oregano, thyme, rosemary, sage, marjoram, peppermint, spearmint, savory, lavender, and basil. See the following page.

Mother-of-thyme (↑), *Thymus praecox*   Spotted dead nettle (↑), *Lamium maculatum*

(←) *Scutellaria scordifolia* is one of several species called skullcap. This species is an ornamental from Asia, but the genus is native throughout the United States.

Chaste tree or chaste berry (→), *Vitex agnus-castus*, was once classified in the verbena family. It has been discovered that several former verbena genera are more closely related to the mints. *Clerodendrum*, *Callicarpa* (beauty berry), and *Tectona* (teak trees) are among others that have been moved to Lamiaceae.

Lamiaceae species illustrated on the opposite page:

Marrubium vulgare L.—horehound
Moluccella laevis L.—bells-of-Ireland, shellflower
Ocimum basilicum L.—sweet basil
Origanum majorana L.—marjoram
Rosmarinus officinalis L.—rosemary
Salvia officinalis L.—sage
Solenostemon scutellarioides (L.) Codd—garden coleus

Rosemary (↑), *Rosmarinus officinalis*, is a shrub with narrow, almost needle-like leaves.

This variety of sweet basil, *Ocimum basilicum* (↑), has relatively smooth, rippled leaves. Basil was domesticated in India. Many culinary varieties have been bred.

Culinary sage, *Salvia officinalis* (↗), has variegated and normal leaf varieties. The leaves are finely wrinkled.

Marjoram (→), *Origanum majorana*, with variegated and normal leaf varieties, is closely related to oregano.

(←) Coleus, *Solenostemon scutellarioides*, is grown for its colorful leaves, which are relatively thin and have sparse hairs. (Information about these plants is listed under *Coleus* and *Plectranthus*, as well as *Solenostemon*.) Leaves of horehound (↙), *Marrubium vulgare*, are relatively thick and are densely covered with hairs. Both species have square stems and show the opposite leaf arrangement of the mint family.

Bells-of-Ireland or shellflower (→), *Moluccella laevis*, has enlarged green calyxes that dwarf its small white flowers. A calyx of fused sepals is characteristic of the mint family, but this species has especially large calyxes.

217

Eudicots

Euasterids I

Lamiales Bromhead

Oleaceae Hoffmannsegg & Link   24 genera/615 species

Genera include:

Chionanthus L.—fringe tree

Forestiera Poiret—swamp privet

*Forsythia Vahl—forsythia, golden bells

*Fraxinus L.—ash tree

Jasminum L.—jasmine

Ligustrum L.—privet

    *Ligustrum lucidum W. T. Aiton—glossy leaf privet, waxleaf privet

Olea L.—olive tree, black ironwood

    *Olea europaea L.—culinary olive tree

Osmanthus Loureiro—osmanthus, devil wood, fragrant olive

Phillyrea L.—mock privet, phillyrea

Syringa L.—lilac

    *Syringa vulgaris L.—common lilac

* illustrated

Oleaceae, The Olive Family

The olive family typically has flowers with two stamens. The members are spring bloomers, whose flowers usually have four petals that are fused into a tube at the base. They vary from bushes with colorful flowers to ash trees with tiny, inconspicuous flowers that lack a perianth. The ash tree is dioecious, but most other species have bisexual flowers. Olive family leaves are opposite and may be simple or odd-pinnately compound.

The spring-flowering bushes of the olive family include:

Common lilac (←), *Syringa vulgaris*

Waxleaf privet (→), *Ligustrum lucidum*

Forsythia (←), *Forsythia* hybrid

This family is named for its oldest cultivated member, the olive. *Olea europaea* (→) has been grown for its oily fruits since prehistoric times. Its two stamens are yellow when the flower first blooms. Later, they turn brown as they mature and release their pollen.

Ash trees, *Fraxinus* species, are dioecious. The staminate flowers (→) are planted more frequently in city landscapes, because the pistillate trees produce abundant samaras. The branch at left has both the current spring's pistillate flowers, which are green and inconspicuous, and the previous year's fruits, the brown samaras.

Eudicots

Euasterids I

Lamiales Bromhead

Orobanchaceae Ventenat   99 genera/2061 species

Genera include:

Agalinis Rafinesque—false foxglove, gerardia

Aureolaria Rafinesque—yellow false foxglove

Castilleja Mutis ex L. f.—Indian paintbrush

    *Castilleja exserta (Heller) Chuang & Heckard—purple owl clover

    *Castilleja integra Gray—orange Indian paintbrush

Cordylanthus Nuttall ex Bentham—bird's-beak, clubflower

Epifagus Nuttall—beechdrops

Euphrasia L.—eyebright

Orobanche L.—broomrape

    *Orobanche cooperi (Gray) Heller—desert broomrape

Pedicularis L.—lousewort, Indian warrior

    *Pedicularis canadensis L.—Canadian lousewort

    *Pedicularis groenlandica Retzius—little elephanthead

Rhinanthus L.—yellow rattle, rattlebox

Striga Loureiro—witchweed

* illustrated

Orobanchaceae, The Broomrape and Paintbrush Family

This family was formerly a part of family Scrophulariaceae. Its members are true parasites that lack chlorophyll or hemiparasites that have chlorophyll, but still draw some of their nutrients from other plants. All have specialized roots, called haustoria, that grow into their host plants' roots and take nutrients. Most have tubular flowers with colorful corollas of various shapes.

The paintbrushes, such as (↓) *Castilleja integra*, have a smaller corolla that is surrounded by larger, colorful floral bracts.

Desert broomrape (↑), *Orobanche cooperi*, lacks chlorophyll. This striking purple plant draws its nutrients from shrubs of the sunflower family, Asteraceae.

Owl clover (←), *Castilleja exserta*, grows in arid areas of the southwestern United States and northwestern Mexico. It usually parasitizes shrubs of Asteraceae.

Red elephant head (↑), *Pedicularis groenlandica*, gets its common name from the shape of its flowers. It grows in wet, mountain meadows in the western United States.

Louseworts have flowers arranged in a spiral pattern. Top view (↑) and side view (→) of Canadian lousewort, *Pedicularis canadensis*.

Eudicots

Euasterids I

Lamiales

221

Eudicots

Euasterids I

Lamiales Bromhead

Plantaginaceae Jussieu   90 genera/1700 species

Genera include:

Antirrhinum L.—snapdragon

    **Antirrhinum majus* L.—garden snapdragon

Asarina Miller—climbing snapdragon, twining snapdragon, asarina

Bacopa Aublet—water hyssop

Besseya Rydberg—besseya, kittentail, coraldrops

Callitriche L.—water starwort

Chelone L.—turtlehead

Collinsia Nuttall—blue-eyed Mary, Chinese houses

Digitalis L.—foxglove

    **Digitalis purpurea* L.—common foxglove

Hebe Commerson ex Jussieu—hebe, New Zealand lilac

Hippuris L.—mare's-tail

Linaria P. Miller—toadflax, linaria, butter-and-eggs

    **Linaria dalmatica* (L.) P. Miller—Dalmatian toadflax

Penstemon Schmidel—penstemon, beardtongue

    *Penstemon hallii Gray—Hall's penstemon

Plantago L.—plantain, Indian wheat, goose tongue

    **Plantago lanceolata* L.—lanceleaf plantain, English plantain, ribwort

    **Plantago patagonica* Jacquin—wooly plantain

Russelia Jacquin—coral plant, firecracker plant

Sairocarpus D. A. Sutton—wild snapdragon

Veronica L.—speedwell, veronica, gypsyweed, neckweed

    *Veronica spicata L.—spiked veronica, garden veronica

* illustrated

Beardtongue (↑),
Penstemon hallii

Snapdragon (↑),
Antirrhinum majus

Dalmatian toadflax (↑),
Linaria dalmatica

Foxgloves (↑),
Digitalis purpurea

Members of this family were formerly included in Scrophulariaceae. Most have showy flowers with tubular corollas. The bump on the lower lip of snapdragon flowers restricts the pollinators that can enter to large, strong bees. Nectar guides, like those in foxgloves, are commonly present. The plantains have inconspicuous flowers with protruding stamens. Veronicas have only two stamens, but most of this family have four to five stamens. The penstemons, commonly called beardtongues, are named for their five stamens, one of which usually has a tuft of hairs instead of an anther.

Veronicas have four petals, but the upper one is actually two fused together. The lower petal is narrower and more pointed than the others. The stamens are reduced to two. *Veronica spicata* (→) is common garden veronica or speedwell. Many other veronicas are grown as ornamentals, including creeping veronicas and mat forms with more open corollas.

Plantains appear to have little in common with the other Plantaginaceae. Their leaves often have parallel veins, (↓) as shown by lanceleaf plantain, *Plantago lanceolata*. This is because the leaves are phyllodes, expanded leaf stems without blades. These common weeds are wind-pollinated and have a reduced perianth. Wooly plantain, *Plantago patagonica* (↓, ↘), has a small, white, four-part corolla.

Eudicots

Euasterids I

Lamiales Bromhead

Scrophulariaceae Jussieu   65 genera/1700 species

Genera include:

Buddleja L.—butterfly bush, summer lilac, orange ball tree

 *Buddleja davidii Franchet—butterfly bush

 **Buddleja marrubiifolia* Bentham—wooly butterfly bush

*Diascia Link & Otto—diascia, twinspur

Leucophyllum Bonpland—silver leaf, Texas Ranger, Chihuahuan sage

Nemesia Ventenat—nemesia

 *Nemesia caerulea Hiern—nemesia

 *Nemesia strumosa Bentham—nemesia

Scropularia L.—figwort, carpenter's square

Verbascum L.—mullein

 **Verbascum thapsus* L.—common mullein

* illustrated

Scrophulariaceae, The Figwort Family

The new family Scrophulariaceae is only a fraction of its former self. Many of the traditional members have moved to families Plantaginaceae and Orobanchaceae. The remaining scrophs have open corollas that are fused at the base. They may have four or five stamens.

Mulleins are widespread weeds, as well as cultivated flowers. Common mullein (↓, ↘), *Verbascum thapsus*, is a biennial with very hairy leaves. It came from Asia, via Europe, and has spread across North America. The flowers have five stamens with yellow hairy filaments.

Butterfly bush (↑), *Buddleja davidii*, has long, tapering inflorescences that are packed with small flowers. It is especially attractive to butterflies. Wooly butterfly bush, *Buddleja marrubiifolia* (↓), is a Texas native that is grown as an ornamental in arid climates.

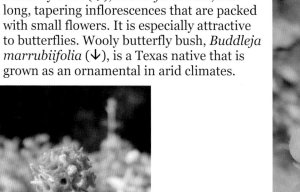

Garden annuals of the Scrophulariaceae

Cultivated nemesias include hybrids of *Nemesia strumosa* (↓) and *N. caerulea* (↓). The twinspurs, *Diascia* hybrids (↓), are named for the two nectar spurs. Both genera are natives of South Africa.

Eudicots

Euasterids I

Lamiales Bromhead

Verbenaceae J. Saint-Hilaire   34 genera/1175 species

Genera include:

Aloysia Jussieu—beebrush, white brush

Duranta L.—golden dewdrops

Glandularia J. F. Gmelin—mock vervain

    **Glandularia bipinnatifida* (Nuttall) Nuttall—Dakota vervain

Lantana L.—lantana, shrub verbena

    *Lantana camara L.—common lantana, bush lantana

    **Lantana urticoides* Hayek—Texas lantana, calico bush

Lippia L.—lippia, Mexican oregano, lemon verbena, fogfruit

Petrea Jussieu

    *Petrea volubilis L.—queen's wreath

Phyla Loureiro—frogfruit, fogfruit, wedgeleaf

*Verbena L.—vervain, verbena

* illustrated

Verbenaceae, The Verbena Family

Dakota vervain (↑), *Glandularia bipinnatifida*

Hybrid variety of bush lantana, *Lantana camara* (↗)

Texas native lantana, *Lantana urticoides* (→)

The verbena family is closely related to the mint family. A number of its traditional members have been transferred to the mints. Verbenas often have square stems and aromatic substances. Their leaves are opposite with a line across the node, and many have serrate margins. The flowers, however, are different from their mint cousins. Many have inflorescences with head-like arrangements. The fused corolla has a narrow tube at the base. It flares out and flattens at the top. Although the corolla is often not perfectly regular, it is not two-lipped and strongly bilateral like a mint flower. The stigma of Verbenaceae flowers is two-lobed and easily seen, unlike the inconspicuous stigmas of the mint family. Verbena family fruits include drupes and schizocarps that split into four nutlets.

Verbena hybrids (→, ↘) are available in many hues for use as bedding plants.

Queen's wreath (←), *Petrea volubilis*, is used in tropical landscaping. Its corolla and sepals are purple. The sepals remain on the developing ovary after the flower has bloomed. Their color gradually fades to green.

Eudicots

Euasterids I

Solanales Dumortier

Convolvulaceae Jussieu   57 genera/1600 species

Genera include:

Calystegia R. Browne—hedge bindweed, false bindweed

> **Calystegia macrostegia* (E. Greene) Brummitt—coast morning glory, island false bindweed

Convolvulus L.—bindweed, bush morning glory

> *Convolvulus arvensis* L.—common bindweed

> **Convolvulus tricolor* L.—dwarf morning glory

Cuscuta L.—dodder

Dichondra J. R. Forster & G. Forster—dichondra

Ipomoea L.—morning glory, sweet potato, moonflower, tievine

> **Ipomoea batatas* (L.) Lamarck—sweet potato

> **Ipomoea leptophylla* Torrey—bush morning glory

> **Ipomoea tricolor* Cavanilles—common morning glory

Jacquemontia Choisy—clustervine, small-flowered morning glory

Merremia Dennstaedt ex Endlicher—Hawaiian wood rose

* illustrated

Convolvulaceae, The Morning Glory Family

Morning glories (↑), hybrids of *Ipomoea tricolor*, are members of the same genus as sweet potatoes. Their fruits are capsules that are covered by the persistent calyx while they develop (↑). The sepals are typically separate or fused only at their bases. When the fruits are mature, the calyx dries and contracts, and the capsule splits and drops its seeds.

The flowers of the morning glory family typically have a fused corolla that is tubular at the base, then flares open like a trumpet. The corolla has five sections and (↑) is fan-folded inside its spiral bud. Many members are vines that have the curious property of always twining clockwise. Dodder, genus *Cuscuta*, looks quite unlike the rest of the family. This parasitic plant appears to be a heap of yellow or orange filaments piled on top of other plants.

Fancy-leaved cultivars of sweet potato (↑), *Ipomaea batatas*, are used as ornamentals.

Coast morning glory (↓), *Calystegia macrostegia*, is a native of California coasts.

Bush morning glory (↑), *Ipomoea leptophylla*, is native to the western plains of the United States. It forms a huge storage root that helps it survive in its arid habitat and gives it its other name, manroot.

Dwarf morning glory (→), *Convolvulus tricolor*, is an annual ornamental plant whose flowers stay open all day. Another member of this genus is the noxious weed, common bindweed, *C. arvensis*. Its deep rhizomes and long-lived seeds make bindweed very difficult to eradicate.

229

Eudicots

Euasterids I

Solanales Dumortier

Solanaceae Jussieu   102 genera/2460 species

Genera include:

Atropa L.—belladonna

Browallia L.—amethyst flower

Brugmansia Persoon—angel's trumpet

Brunfelsia L.—yesterday-today-and-tomorrow

Calibrachoa Cervantes—million bells, miniature petunia

Capsicum L.—peppers, chili pepper, bell pepper

Cestrum L.—night jasmine

Datura L.—jimson weed, datura, thorn-apple

    *Datura innoxia Miller—sacred datura

Hyoscyamus L.—henbane

Lycium L.—desert thorn, wolfberry

Lycopersicon, the garden tomato, is now included in *Solanum*

Mandragora L.—mandrake

Nicotiana L.—tobacco, nicotiana

    *Nicotiana alata Link & Otto—flowering tobacco, nicotiana

Petunia Jussieu—petunia

    *Petunia x *hybrida* (Hooker) Vilmorin—bedding petunia

Physalis L.—ground cherry, husk tomato, tomatillo, Chinese lantern

Quincula Rafinesque—Chinese lantern, purple ground cherry

Salpiglossis Ruiz & Pavón—salpiglossis, painted tongue

Schizanthus Ruiz & Pavón—schizanthus, poor man's orchid, butterfly flower

Solanum L.—eggplant, potato, garden tomato, nightshade, horse nettle, potato creeper

    **Solanum dulcamara* L.—bittersweet nightshade

    **Solanum elaeagnifolium* Cavanilles—silverleaf nightshade, purple nightshade

    **Solanum rostratum* Dunal—buffalo bur, Kansas thistle, mal mujera

* illustrated

Solanaceae, The Nightshade Family

Petunias (→), *Petunia* x *hybrida*, and flowering tobacco (←), *Nicotiana alata*, are popular garden annuals of the nightshade family.

The nightshade family is notorious for its alkaloids, molecules that are poisonous or used as medicines. In spite of this, family members include important food plants (see the following page). The flowers have a fused corolla that is usually is open and flat or even folded back toward the stem. The stamens are grouped in the center around the single style and may stick together. The two-carpellate ovary develops into a berry, a capsule, or a schizocarp of nutlets. The calyx persists after the flower blooms and often enlarges with the fruit. In the tomatillo, genus *Physalis*, and its wild relatives, the calyx completely covers the developing fruit.

Silverleaf nightshade, also called purple nightshade, *Solanum elaeagnifolium* (←), has fruits that look like tiny golden tomatoes, but they are hard and dry, and often turn dark as they age.

Buffalo bur (→), *Solanum rostratum*, has stout prickles on its leaves and stems. A prickly calyx covers the developing fruit.

Bittersweet nightshade (←), *Solanum dulcamara*, is a Eurasian native that has become naturalized throughout much of North America. The bright red, ripened fruits are mildly poisonous. The green fruits are more poisonous than the ripe ones.

Sacred datura, *Datura innoxia* (↓), is also called thornapple, because of its spiny-looking fruit. The fruit shown below (↘) has the remains of its dried corolla, which has been pushed off the end of the developing fruit.

Solanaceae species illustrated on the opposite page:

Capsicum annuum L.—bell pepper, wax pepper, chili pepper, jalapeño, pimento
Physalis ixocarpa Brotero ex Hornemann—tomatillo
Solanum lycopersicum L.—garden tomato
Solanum melongena L.—eggplant, aubergine
Solanum tuberosum L.—potato

More Solanaceae: Domesticated Species Used for Food

Tomatoes (↓), *Solanum lycopersicum*, bear fruits that are wonderfully edible, but the leaves and stems are poisonous. Botanically, tomatoes are fruits, not vegetables. Botanically, tomato fruits are berries.

Eggplants (←), *Solanum melongena*, have purple flowers that closely resemble other nightshades. The persistent calyx caps the stem end of the fruit.

A wide variety of peppers have been bred from the species, *Capsicum annuum* (→) including both sweet and hot peppers. Chili peppers have more capsaicin, the substance that tastes spicy. The intensely spicy tabasco and habenero peppers also belong to this genus.

The fruits of tomatillos (↓), *Physalis ixocarpa*, develop within the calyx, which enlarges along with the growing fruits.

The flowers of the potato, *Solanum tuberosum,* do not produce edible fruit (↓). The edible parts of potatoes are the tubers, underground stems that are modified to store food reserves.

Eudicots

Euasterids II

Aquifoliales Senft

Aquifoliaceae A. Richard   1 genus/405 species

Genus:

Ilex L.—holly, yerba maté, yaupon, ink berry, winterberry

*Ilex aquifolium L.—common holly, European holly

*Ilex x *meserveae* S. Y. Hu—blue holly

*Ilex opaca Aiton—American holly

* illustrated

Aquifoliaceae, The Holly Family

Ilex is the sole genus of this family of shrubs and trees. Most hollies have spiny leaf margins, but leaves with smooth margins also occur. Hollies are often evergreen, with shiny, leathery leaves. The flowers usually have four to six petals that are fused at the base. There is no little or no style; the stigma is attached directly to the ovary. Hollies are usually dioecious, although there are a few cultivars with bisexual flowers. In the pistillate flowers, there are staminodes, nonfunctional stamen-like structures. Likewise, the staminate flowers often have a remnant of a pistil. The fruit is a drupe with several stones. Hollies are native to eastern North America and occur in much of the rest of the world.

The pistillate flowers (↑) of holly clearly show the superior ovary in this *Ilex aquifolium* cultivar. It has staminodes between the petals. The staminate flower of blue holly, *Ilex* x *meserveae* (↑), has four functional stamens that are attached to the petals.

Cultivars of American holly (↑), *Ilex opaca*, and blue holly (↑), *I.* x *meserveae*.

The fruits of hollies are drupes that ripen to red or black and that usually contain four to six stones. Each stone holds a single seed. The number of stones shows the number of carpels in the pistil. There is a dark disk (↑) on the end of the fruit, which is the remnant of the prominent stigma. Holly seeds often take a few years to germinate.

Eudicots

Euasterids II

Apiales Nakai

Apiaceae Lindley or Umbelliferae Jussieu   434 genera/3780 species

Genera include:

Aegopodium L.—bishop's weed, goutweed

Aletes J. M. Coulter & Rose—mountain caraway, Indian parsley

Anethum L.

    **Anethum graveolens* L.—dill

Angelica L.—angelica

Anthriscus Persoon—chervil

Apium L.—celery

Astrantia L.—masterwort

Berula W. D. J. Koch—water parsnip

Bupleurum L.—thoroughwax

Carum L.—caraway

Cicuta L.—water hemlock

Coriandrum L.—coriander

Cuminum L.—cumin

Daucus L.—carrot

    *Daucus carota L.—carrot, wild carrot, Queen Anne's lace

*Eryngium L.—sea holly, rattlesnake master, eryngo, coyote thistle

Foeniculum Miller—fennel

    **Foeniculum vulgare* Miller—sweet fennel, wild fennel, Florence fennel

Glehnia F. Schmidt ex Miquel— beach silvertop

Heracleum L.—cow parsnip, giant hogweed

Levisticum Hill—lovage

Ligusticum L.—licorice root, lovage

Lomatium Rafinesque—biscuit root, desert parsley

Myrrhis Miller

    *Myrrhis odorata (L.) Scopoli—sweet cicely

Oreoxis Rafinesque—alpine parsley, oreoxis

Osmorhiza Rafinesque—wild sweet cicely

Pastinaca L.—parsnip

Perideridia Reichenbach—yampa

Petroselinum Hill—parsley

    **Petroselinum crispum* (Miller) Nyman ex A. W. Hill—parsley, parsley root

Pimpinella L.—anise

Sanicula L.—black snackroot

Sium L.—water parsnip

Zizia W. D. J. Koch—golden alexanders

* illustrated

Apiaceae or Umbelliferae, The Carrot and Parsley Family

The members of Apiaceae can be recognized by the compound umbels of tiny flowers and the sheathing bases of their leaf stems. The leaves of most species are once or twice pinnately compound. The flowers have a two-carpellate pistil that bears two separate styles. The fruits are schizocarps that split in two.

The Apiaceae are aromatic and supply us with many foods and seasonings. Several species have been domesticated since prehistoric times. Note that we eat the sheathing petioles of celery and fennel, not the stems. This family also has many highly poisonous members. One should **never eat or taste wild species**. Poisonous genera include *Conium,* poison hemlock, thought to be the plant used to execute Socrates and *Cicuta,* water hemlock, which is deadly poisonous even in small amounts.

The leaves of parsley, *Petroselinum crispum* (←), show the sheathing petiole that is characteristic of Apiaceae (white arrow). Parsley, like carrots, is a biennial. The first year it is a rosette plant, with little visible stem. The leaf attachment is most easily seen on the second-year, flowering stem, as shown here.

Sweet cicely (↑), *Myrrhis odorata*, shows the typical compound leaves of this family. Its developing fruits (↑) retain the pistil's two styles.

Dill (←), *Anethum graveolens*, shows the typical form of the inflorescence, the compound umbel.

The view from underneath an inflorescence of Queen Anne's lace or wild carrot (→), *Daucus carota*, shows the structure of the small umbels that are joined into the large, compound umbel.

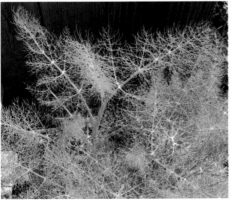

Sea holly (→), *Eryngium*, does not look like the typical carrot family member. It has a reduced, cone-shaped inflorescence and thistle-like foliage.

Fennel (←), *Foeniculum vulgare*, grows wild in many parts of the world. Its leaf blades are finely divided into many linear segments.

Eudicots

Euasterids II

Asterales Lindley

      Asteraceae Martynov or **Compositae** Giseke   1528 genera/22750 species

      (Genera are listed on the following pages under the individual tribes and in the outline on pages 274-275.)

The Diversity of the Sunflower Family

On this tour, we follow the flowering plant family and subfamily structure of the Angiosperm Phylogeny Group classification. The sunflower family, however, is so large and holds such enormous diversity that its subfamilies need further subdivision into tribes. Traditional tribes are clearly not natural groups—they have a mix of ancestries. Our understanding of relationships within the sunflower family is still developing, but we will look to the future, even if that picture is still subject to change. The source for the following Asteraceae tribes is the findings of Vicki Funk and coworkers (Funk et al., 2005, listed in this book's references).

To give you an idea of the diversity of Asteraceae, there are currently at least 10 subfamilies plus additional groups proposed for Asteraceae. We will visit the three subfamilies that are native to temperate North America and found worldwide.

This tour visits the following groups:

Thistle subfamily, Carduoideae
        Cardueae, the thistle tribe

Chicory subfamily, Cichorioideae
        Cichorieae, the lettuce tribe

Aster subfamily, Asteroideae
        Anthemideae, the anthemis tribe
        Astereae, the aster tribe
        Coreopsideae, the coreopsis tribe
        Eupatorieae, the gayfeather tribe
        Gnaphalieae, the everlastings tribe
        Helenieae, the sneezeweed tribe
        Heliantheae, the sunflower tribe
        Senecioneae, the senecio tribe

Asteraceae or Compositae, The Sunflower Family

Asteraceae flowers form in inflorescences called heads. The heads, which appear to be single flowers, are composites of disk flowers, ray flowers, or both. At first glance, ray flowers look like petals, but a closer look reveals a tiny tube at the base, in some cases with a style inside. Disk flowers have small tubular corollas, typically with five equal points at the top of the tube. The disk flowers are usually bisexual and fertile. A few species have separate staminate and pistillate flowers. Most ray flowers are either pistillate or sterile, in which case they serve only to attract pollinators. The lettuce tribe has heads composed only of bisexual ray flowers.

The anthers of the five stamens fuse together and form a tube through which the style grows, carrying the pollen upward. At maturity, the two style branches curl open, revealing the stigmas. These pollen-receiving surfaces are reduced to lines on the style branches. The ovary is inferior and has a single ovule.

The whole inflorescence has an involucre of bracts at its base. These bracts are called phyllaries, and they are often mistaken for sepals. The true sepals in each flower are reduced to bristles, scales, or awns and are called the pappus. The fruit is an achene, which may be crowned by the dried pappus at maturity.

Tribes differ in the structure of the phyllaries and in whether there are one or more rows. A few species lack the involucre. The nature of the pappus and whether it is present are important characters for identifying tribes. The heads may be single like sunflowers or grouped together into larger, secondary inflorescences, as seen in goldenrod and yarrow.

(←) A cross section of the sunflower, *Helianthus annuus*, shows that it is really an inflorescence whose many small flowers sit on a wide receptacle.

There are two kinds of flowers in the inflorescence. Ray flowers ring the outside. They look superficially like petals, but each is a flower with a tubular corolla that opens and extends to one side. The center of the inflorescence is covered with disk flowers. The disk flowers around the periphery bloom first. The center disk flowers are the youngest and last to bloom.

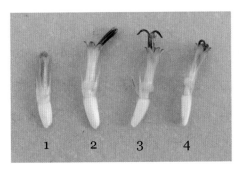

(←) Individual disk flowers from a sunflower, youngest to oldest shown left to right, each with their white, inferior ovary. 1) The flower in bud. 2) A dark tube of fused anthers protrudes from the corolla. 3) The style has emerged through the anther tube and its two branches have spread apart. 4) The style and anther tube wither as the flower finishes blooming. The corolla will remain on the developing ovary. The two pointed scales atop each ovary at the base of the corolla are the pappus, the modified sepals.

An enlargement of the disk flowers of the green coneflower, *Rudbeckia ampla* (→), shows the black tube formed by the fused anthers of the flowers near the top. Just outside these, on older flowers, the yellow styles project past the anthers, and their two branches have opened (arrow). The stigmas are simply lines on the interior surface of the style branches.

239

Eudicots

Euasterids II

Asterales Lindley

Asteraceae Martynov or Compositae Giseke

Subfamily Carduoideae Cassini ex Sweet: Tribe Cardueae Cassini

Genera include:

Arctium L.—burdock

Carduus L.—thistle

> **Carduus nutans* L.—musk thistle, bristle thistle

Carthamus L.—safflower, saffron thistle

Centaurea L.—cornflower, bachelor's buttons, knapweed

> **Centaurea biebersteinii* DC.—spotted knapweed

> **Centaurea montana* L.—perennial cornflower

Cirsium Miller—thistle

> **Cirsium undulatum* (Nutall) Sprengel—wavyleaf thistle

Cynara L.—artichoke

> **Cynara cardunculus* L.—cardoon

Onopordum L.—Scotch thistle

Silybum Adanson—milk thistle

Xeranthemum L.—immortelle

* illustrated

The thistle tribe includes thistles, artichokes, and cornflowers. The typical disk flower of this tribe has a deeply lobed corolla and a long style, which make the heads look very fluffy. Ray flowers are usually absent, although cornflowers have tubular, sterile ray flowers. There are several rows of phyllaries in the involucre. Each phyllary ends in a sharp spine or is edged with teeth. The burdocks, *Arctium*, have phyllaries that end in a long, slender hook.

Disk flowers of bristle thistle (↑), *Carduus nutans*, show their deep corolla lobes (arrows) and their white pappus, a modified calyx. The pappus remains on the fruit and aids in its dispersal by wind. The style branches of many tribe members are very small and not easily seen without magnification. The style itself projects well past the corolla.

Wavyleaf thistle (←), *Cirsium undulatum*, has an involucre with several rows of bracts, each with a sharp spine.

Thistles are often a symbol of endurance and toughness. Although native thistles are usually not invasive, introduced species are often noxious weeds. Spotted knapweed, *Centaurea biebersteinii* (→), is an example. It was accidentally introduced from Eurasia and readily displaces North American native plants.

Perennial cornflower (↑), *Centaurea montana*, is a common garden flower.

The cardoon, *Cynara cardunculus* (↗), is an artichoke relative, which is grown as an ornamental and for its edible leaf stalks. When we eat an artichoke, we are consuming the starchy food reserve in the bases of the phyllaries and in the receptacle of the inflorescence.

Eudicots

Euasterids II

Asterales

Eudicots

Euasterids II

Asterales Lindley

Asteraceae Martynov or **Compositae** Giseke

Subfamily Cichorioideae (Cassini) Chevallier: Tribe Cichorieae Lamarck & DC.

Genera include:

Agoseris Rafinesque—mountain dandelion

Catananche L.—cupid's-dart

Chondrilla L.—hog bite, skeletonweed

Cichorium L.—endive, chicory, radicchio, escarole

Cichorium intybus L.—chicory

Crepis L.—hawksbeard

Hieracium L.—hawkweed, rattlesnake weed

Krigia Schreber—dwarf dandelion

Lactuca L.—lettuce, prickly lettuce, wild lettuce

Lactuca sativa L.—garden lettuce

Lygodesmia D. Don—skeletonplant

Prenanthes L.—rattlesnake root, white lettuce, gall-of-the-earth

Pyrrhopappus DC.—Texas dandelion, false dandelion, desert chicory

Sonchus L.—sowthistle

Taraxacum Weber ex Wiggers—dandelion

Taraxacum officinale Weber ex Wiggers—common dandelion

Tragopogon L.—salsify, oyster root, goatsbeard

Tragopogon dubius Scopoli—yellow salsify, goatsbeard

* illustrated

Asteraceae: Subfamily Cichorioideae—Cichorieae, The Lettuce Tribe

Heads of the lettuce tribe have only ray flowers; these are bisexual and typically have five tiny points at the end of the corolla. They are called ligulate flowers to distinguish them from other ray flowers in Asteraceae, which are pistillate or sterile and typically have corollas with three points. The pappus is like a little umbrella that aids in wind dispersal of the achenes. In many species, it is set on a stem-like projection of the seed that is called a beak. Plants of this tribe have milky sap, a type of latex.

Yellow salsify (↑), *Tragopogon dubius*, has phyllaries that project beyond its yellow ray flowers. These involucral bracts close back over the developing fruits after the flowers have bloomed. When the fruits are mature, the phyllaries reopen. (↑) Each fruit has a long stem-like beak, topped by its fluffy pappus.

Chicory (↑), *Cichorium intybus*, is a common roadside weed. It is a Mediterranean native that is sometimes cultivated for its roots, which are used as a coffee flavoring.

The common dandelion, *Taraxacum officinale* (↑), is a widespread weed in temperate climates. It is a hybrid that produces its abundant seeds asexually.

Garden lettuce (→), *Lactuca sativa*, has been cultivated for thousands of years. There are many cultivated varieties with different leaf characteristics. The flower heads form on a tall stem that arises from the basal rosette of leaves. Our garden varieties have much less of the bitter latex than wild lettuces.

Eudicots

Euasterids II

Asterales Lindley

Asteraceae Martynov or Compositae Giseke

Subfamily Asteroideae Lindley: Tribe Anthemideae Cassini

Genera include:

Achillea L.—yarrow
    **Achillea lanulosa* Nuttall—native yarrow, plumajillo
    **Achillea millefolium* L.—common yarrow, milfoil
Anacyclus L.—Atlas daisy
Anthemis L.—anthemis, dog fennel, mayweed
    **Anthemis tinctoria* L.—golden marguerite
Artemisia L.—sagebrush, sagewort, wormwood, tarragon, southernwood, mugwort
    **Artemisia arctica* Lessing—arctic sagewort
Chamaemelum Miller—chamomile (ground cover)
Chrysanthemum L.—painted daisy, marguerite, chrysanthemum
    **Chrysanthemum morifolium* Ramatuelle—florist's chrysanthemum
Leucanthemum Miller—Shasta daisy, oxeye daisy
Matricaria L.—chamomile tea, matricaria
    **Matricaria discoidea* DC.—pineapple weed
Santolina L.—lavender cotton, santolina
Tanacetum L.—tansy, feverfew, pyrethium

* illustrated

The heads of the anthemis tribe have ray and disk flowers or disk flowers alone. The leaves are often finely divided (→) and are typically aromatic. There are several rows of phyllaries. The pappus, if it is present, is scaly. The fruits do not bear the bristly pappus found in many other tribes.

Pineapple weed (↑), *Matricaria discoidea*, grows worldwide. If you pick an inflorescence and crush it between your fingers, it releases a strong, pineapple-like fragrance.

Golden marguerite (↑), *Anthemis tinctoria*, is a common garden flower that has been used as a dye plant.

Chrysanthemums (→), *Chrysanthemum morifolium*, are used as cut flowers by the florist industry. They are available in many colors and with many shapes of ray flowers.

Members of the genus *Achillea*, the yarrows, include both cultivated and native plants. The umbel-like inflorescences are groups of many small heads. A cultivar of *A. millefolium* has red ray flowers (↓). Native yarrow (↓), *A. lanulosa*, grows throughout western North America.

Arctic sagewort (→), *Artemisia arctica*, lives in the alpine tundra. Other members of this genus include woody shrubs and herbaceous plants of temperate climates.

245

Eudicots

Euasterids II

Asterales Lindley

> **Asteraceae** Martynov or **Compositae** Giseke
>
> Subfamily Asteroideae Lindley: **Tribe Astereae** Cassini
>
>> Genera include:
>>
>> *Aster* L.—aster, Michaelmas daisy
>>
>> *Bellis* L.—English daisy
>>
>> *Boltonia* L'Héritier—doll's daisy
>>
>> *Brachycome* Cassini—Swan River daisy
>>
>> *Callistephus* Cassini—China aster, callistephus
>>
>> *Chrysopsis* (Nuttall) Elliot—golden aster
>>
>> *Chrysothamnus* Nuttall—see *Ericameria*
>>
>> *Conyza* Lessing—horseweed, asthmaweed
>>
>>> **Conyza canadensis* (L.) Cronquist—common horseweed
>>
>> *Ericameria* Nuttall—goldenbush, rabbitbrush
>>
>>> **Ericameria nauseosa* (Pallas ex Pursh) G. L. Nesom & G. I. Baird—
>>> rubber rabbitbrush, gray rabbitbrush, chamisa
>>
>> *Erigeron* L.—erigeron, fleabane
>>
>>> **Erigeron speciosus* (Lindley) DC.—aspen fleabane
>>
>> *Felicia* Cassini—blue marguerite, kingfisher daisy
>>
>> *Grindelia* Willdenow—gumweed
>>
>>> **Grindelia subalpina* Greene—subalpine gumweed
>>
>> *Heterotheca* Cassini—golden aster, false golden aster, telegraph weed, camphor weed
>>
>>> **Heterotheca villosa* (Pursh) Shinners—hairy golden aster
>>
>> *Machaeranthera* Nees—tansy aster
>>
>>> **Machaeranthera tanacetifolia* (Kunth) Nees—tansyleaf tansy aster
>>
>> *Solidago* L.—goldenrod
>>
>>> **Solidago canadensis* L.—Canada goldenrod
>>
>> *Townsendia* Hooker—Easter daisy
>>
>>> **Townsendia hookeri* Beaman—Hooker's Easter daisy

* illustrated

Tansy aster (↑), *Machaeranthera tanacetifolia*, grows throughout the western United States.
This hairy golden aster, *Heterotheca villosa* (↑), has mature fruits (left) with a pappus of bristles.
Aspen fleabane, *Erigeron speciosus*, shows the many narrow ray flowers typical of its genus (↑).

Heads of the aster tribe usually have both ray and disk flowers, although a few members have only disk flowers. There are several rows of phyllaries in the involucre. The typical pappus is a tuft of bristles on the fruit. The Astereae is a large tribe, with nearly 3000 species. The genus *Aster* and other large genera have been broken up into smaller genera that better reflect our understanding of their relatedness.

Subalpine gumweed (↑), *Grindelia subalpina*, has curled phyllaries and a white, gummy substance on its immature heads (arrow), typical of the genus.
Canada goldenrod, *Solidago canadensis* (↑), grows throughout the United States and Canada.
Rubber rabbitbrush, *Ericameria nauseosa*, is a shrub of dry, sunny areas; it has only disk flowers (↑).

Horseweed (←), *Conyza canadensis*, is a common weed in much of North America. Its tiny flowers are barely visible above their green involucres.

Hooker's Easter daisy, *Townsendia hookeri* (→), blooms in early spring, in contrast to many other Astereae, which bloom in late summer.

Eudicots

Euasterids II

Asterales

247

Eudicots

Euasterids II

Asterales Lindley

Asteraceae Martynov or **Compositae** Giseke

Subfamily Asteroideae Lindley: Tribe Coreopsideae Lindley

Genera include:

Bidens L.—beggar-ticks, sticktight, Spanish needle

*Bidens tenuisecta Gray—slimlobe sticktight

Coreopsis L.—tickseed

*Coreopsis verticillata L.—threadleaf coreopsis, whorled tickseed

Cosmos Cavanilles—cosmos

*Cosmos bipinnatus Cavanilles—garden cosmos

*Dahlia Cavanilles—dahlia

* illustrated

Many of the coreopsis tribe are cultivated as ornamentals. They include dahlias (←), *Dahlia*, (↑) threadleaf coreopsis, *Coreopsis verticillata*, and (↓) cosmos, *Cosmos bipinnatus*.

The coreopsis tribe is one of 12 that have been created recently from the former sunflower tribe. The involucre usually has two rows of phyllaries that differ in shape. The inner row lies closely against the ray flowers, whereas the outer row often looks more like leaves and stands out from the stem. A cosmos shows the typical involucre (↘). The pappus of several tribe members consists of two to four barbed awns that catch in fur or clothing and help in seed dispersal. The common names of tickseed and beggar-ticks reflect this mode of seed travel.

(↓) Fruits of slimlobe sticktight, *Bidens tenuisecta*, each have two barbed awns on the top. (↓) When the fruits are mature, they fan out and present their clingy awns to passing animals.

Eudicots

Euasterids II

Asterales

Eudicots

Euasterids II

Asterales Lindley

Asteraceae Martynov or Compositae Giseke

Subfamily Asteroideae Lindley: Tribe Eupatorieae Cassini

Genera include:

Ageratina Spach—snakeroot

Ageratum L.—whiteweed

    *Ageratum houstonianum Miller—floss flower, ageratum

Brickellia Elliott—brickelbush

Carphephorus Cassini—chaffhead, vanillaleaf

Chromolaena DC.—thoroughwort

Conoclinium DC.—blue mistflower, thoroughwort

Eupatorium L.—Joe-Pye weed, boneset, thoroughwort, snakeroot

    *Eupatorium maculatum L.—spotted joe-pye weed

Liatris Gaertner—gayfeather, liatris, blazingstar

    *Liatris punctata Hooker—dotted blazing star, dotted gayfeather

Mikania Willdenow—climbing boneset, climbing hempweed

Stevia Cavanilles—stevia, candyleaf

* illustrated

250

(↑) Ageratum or floss flower, *Ageratum houstonianum*, is a common garden flower. There are many cultivars, including this one (←), which has white corollas and blue style appendages.

Spotted Joe-pye weed, *Eupatorium maculatum* (↑), along with several related species, is common in eastern North America. This species also grows in the northwest and is the only *Eupatorium* native species that lives west of the Rocky Mountains.

Heads of the gayfeather tribe have only disk flowers. A notable feature of this tribe is the style structure. The two style branches are long and thread-like, which gives the flowers a feathery look. The corollas are white or shades of red to blue, but not yellow. The involucre usually has two or more rows of phyllaries.

Dotted gayfeather, *Liatris punctata* (→), blooms in late summer to fall. It lives in the midcontinent of North America, usually in dry grasslands or shrub lands. Other members of this genus grow in moist lowlands. The genus *Liatris* is found from the eastern foothills of the Rocky Mountains to the east coast.

Eudicots

Euasterids II

Asterales

251

Eudicots

Euasterids II

Asterales Lindley

Asteraceae Martynov or Compositae Giseke

Subfamily Asteroideae Lindley: Tribe Gnaphalieae Cassini ex Lecoq & Juillet

Genera include:

Anaphalis DC.—pearly everlasting, anaphalis
> **Anaphalis margaritacea* (L.) Bentham—western pearly everlasting

*Antennaria Gaertner—pussytoes
> **Antennaria neglecta* Greene—field pussytoes

Craspedia G. Forster—drumsticks

Filago L.—cottonrose

Gnaphalium L.—cudweed

Helichrysum Miller—strawflower, curry plant
> **Helichrysum bracteatum* (Ventenat) Anderberg—annual strawflower

Leontopodium R. Browne ex Cassini
> **Leontopodium alpinum* Cassini—edelweiss

Pseudognaphalium Kirpicznikov—cudweed, rabbit-tobacco

Stylocline Nuttall—neststraw

* illustrated

252

The heads of the everlasting tribe usually have only disk flowers. The phyllaries of many species are thin, papery, and white or brightly colored. They persist long after the flowers bloom and are the feature for which the everlastings are named. The pappus is typically a tuft of fine bristles. The leaves of this tribe have entire margins.

This pearly everlasting (←), *Anaphalis margaritacea*, has finished blooming and its seeds are maturing in the brown areas. The white structures are its phyllaries.

Strawflower (↑), *Helichrysum bracteatum*, is grown for its colored, papery phyllaries, which are pink in this example. For floral arrangements, its heads are usually picked and dried before the flowers bloom.

(↑) Edelweiss, *Leontopodium alpinum*, has a conspicuous whorl of fuzzy white leaves beneath its group of flower heads. The individual disk flowers are quite small.

(←) Pussytoes, *Antennaria*, forms low mats of foliage from which the bloom stalks arise. It can be difficult to identify to species because natural hybrids often occur. Some colonies have only pistillate flowers and reproduce asexually. Field pussytoes, *Antennaria neglecta*, has both staminate (↓) and pistillate (↓) flowers on separate plants.

Eudicots

Euasterids II

Asterales

253

Eudicots

Euasterids II

Asterales Lindley

Asteraceae Martynov or Compositae Giseke

Subfamily Asteroideae Lindley: Tribe Helenieae Lindley

Genera include:

Baileya Harvey & A. Gray ex Torrey—desert marigold

Balduina Nuttall—honeycomb head

Gaillardia Fougeroux—gaillardia, Indian blanket, blanket flower

> **Gaillardia* x *grandiflora* Van Houtte—gaillardia, blanket flower

Helenium L.—sneezeweed, yellowdicks

> **Helenium autumnale* L.—common sneezeweed

Hymenoxys Cassini—bitterweed, rubberweed

Marshallia Schreber—Barbara's buttons

Psilostrophe DC.—paperflower

Tetraneuris Greene—alpine sunflowers, alpine bitter weed, perky Sue, four-nerve daisy

> **Tetraneuris acaulis* (Pursh) Greene—goldflower, butte marigold, Angelita daisy

> **Tetraneuris grandiflora* (Torrey & A. Gray) Greene—alpine sunflower, old-man-of-the-mountain

* illustrated

254

Alpine sunflower (→) or old-man-of-the-mountain, *Tetraneuris grandiflora*, lives in the extreme environment of the alpine tundra. It is often accompanied by the related goldflower (↑), *Tetraneuris acaulis*, a more compact species that also grows at lower elevations.

The sneezeweed tribe can be hard to define on the basis of its structures. It has pale anthers, as opposed to the dark ones of the sunflower tribe. The ray flowers are yellow to red, not blue or purple. Their corollas are often wedge-shaped, broader at the apex, and narrow at the base.

The style branches of this sneezeweed (←), *Helenium autumnale*, make a pattern of yellow lines against the darker corollas of the disk flowers.

Gaillardia or blanket flower (←, ↓), *Gaillardia* x *grandiflora*, was bred from two species that are native throughout the United States.

255

Eudicots

Euasterids II

Asterales Lindley

Asteraceae Martynov or Compositae Giseke

Subfamily Asteroideae Lindley: Tribe Heliantheae Cassini

Genera include:

Ambrosia L.—ragweed, bursage

    **Ambrosia psilostachya* DC.—western ragweed

Balsamorhiza Hooker ex Nuttall—balsamroot

Berlandiera DC.—greeneyes

    **Berlandiera lyrata* Bentham—chocolate daisy, lyreleaf green eyes

Echinacea Moench—purple coneflower

Helianthella Torrey & A. Gray—dwarf sunflower, helianthella

Helianthus L.—sunflower, Jerusalem artichoke

    **Helianthus annuus* L.—common sunflower

Iva L.—marshelder, sumpweed

Ratibida Rafinesque—prairie coneflower

    **Ratibida columnifera* (Nuttall) Wooton & Standley—Mexican hat

Rudbeckia L.—blackeyed Susan, browneyed Susan, coneflower

    *Rudbeckia ampla Nelson—green coneflower, golden glow, tall coneflower
        (illustration on page 239)

Silphium L.—rosinweed, compass plant

Tithonia Desfontaines ex Jussieu—Mexican sunflower

Wyethia Nuttall—mule-ears

Xanthium L.—cocklebur

Zinnia L.—zinnia

    **Zinnia grandiflora* Nuttall—plains zinnia, wild zinnia

* illustrated

Heads of the sunflower tribe usually have both ray and disk flowers, although some species lack ray flowers. The pappus may be scales, awns, or bristles. The involucre has more than one row of phyllaries (↓). The anthers of this tribe are often black or dark-colored. In sunflowers, the ray flowers are usually sterile; some genera have pistillate ray flowers. The corollas (↓) of the disk flowers remain on the developing fruits of sunflowers. The fruits, which are achenes, are often called sunflower seeds, but the actual seed is inside the thin, brittle hull.

Western ragweed (←), *Ambrosia psilostachya*, has tiny, wind-pollinated flowers. It is monoecious, with many staminate flower heads borne at the branch ends. The pistillate flowers (↑) in the leaf axils below are much less conspicuous. They lack both a corolla and an involucre. The abundant pollen of the ragweeds causes allergies in many people.

The plains zinnia (↓), *Zinnia grandiflora*, (↓) the prairie coneflower or Mexican hat, *Ratibida columnifera*, and the chocolate flower or green eyes, *Berlandiera lyrata* (↓), are three of the many species of Heliantheae that are native in the United States.

Eudicots

Euasterids II

Asterales

Eudicots

Euasterids II

Asterales Lindley

Asteraceae Martynov or Compositae Giseke

Subfamily Asteroideae Lindley: Tribe Senecioneae Cassini

Genera include:

Arnoglossum Rafinesque—Indian plantain, cacalia

Cineraria L.—cineraria (yellow flower)

Doronicum L.—leopard's bane

Erechtites Rafinesque—burnweed

Ligularia Cassini—ligularia

    **Ligularia wilsoniana* (Hemsley) Greenman—giant groundsel

Packera Á. Löve & D. Löve—groundsel, ragwort, butterweed

Pericallis D. Don

    **Pericallis hybrida* B. Nordenstam—florist's cineraria, hybrid cineraria

Petasites Miller—coltsfoot, butterbur

Senecio L.—senecio, groundsel, ragwort

    **Senecio cineraria* DC.—dusty miller

    **Senecio integerrimus* Nuttall—lambstongue groundsel

    **Senecio rowleyanus* Jacobsen—string-of-beads

Tetradymia DC.—horsebrush, cottonthorn

Tussilago L.

    **Tussilago farfara* L.—coltsfoot

* illustrated

Finding a Good Place for the Bogbeans

The bogbean or buckbean family, Menyanthaceae, has had many different classifications. Aquatic plants can be difficult to classify because their form is often altered greatly as they adapt to a water habitat. The bogbeans were an especially difficult puzzle. This family of aquatic plants was first classified as a subfamily of the gentian family. Then it was placed in its own family, which has been put within a phlox order, a nightshade order, and a campanula order. It was even placed in its own separate order.

Finally, the DNA evidence showed the closest relatives of the Menyanthaceae. The bogbeans clearly belong within the Asterales. This result was a surprise to botanists. The bogbeans do not show plunger pollination (see page 260), one of the notable features of most Asterales. They probably evolved from an ancestor with plunger pollination and then lost that feature. They have a superior ovary, unlike the two main families in the order, Asteraceae and Campanulaceae. In spite of this, they share many other features of plant chemistry and structure with the order.

Nymphoides peltata (←) has fringes on its petals, which are characteristic of the Menyanthaceae.

Lambstongue groundsel (↑), *Senecio integerrimus*, has the typical involucre. Wilson's ligularia (↑), *Ligularia wilsoniana*, florist's cineraria (↑), *Pericallis hybrida*, and coltsfoot (↓), *Tussilago farfara*, are cultivated as ornamentals.

The flowers of this tribe have a single row of phyllaries that are all the same length. The involucre is roughly cylindrical, and the phyllaries are frequently fused. The heads can have both disk and ray flowers or only disk flowers. The pappus consists of a thatch of fine white bristles. The genus *Senecio* is quite large, with over a thousand species. Its members are adapted to environments from deserts to alpine slopes. There are many common names for its members, including groundsel, ragwort, and butterweed.

Senecio cineraria (→), is one of several species with woolly gray-green leaves that bear the common name of dusty miller. It is a perennial in climates with mild winters, but it is grown as an annual elsewhere.

String-of-beads (←), *Senecio rowleyanus*, is an African succulent that is grown as a houseplant. The green, bead-like structures are its leaves, which are modified to store water. Worldwide within the Senecioneae, there are vining species that look like ivy, shrubs, tree-like forms, and many succulents, as well as the more familiar herbaceous forms.

Eudicots

Euasterids II

Asterales Lindley

Campanulaceae Jussieu   79 genera/2300 species

Genera include:

Adenophora Fisher—ladybell

Campanula L.—campanula, bellflower, harebell

     *Campanula glomerata L.—clustered bellflower

     *Campanula rotundifolia L.—harebell, bluebell of Scotland

     *Campanula takesimana Nakai—Korean bellflower

Codonopsis Wallich—bonnet bellflower

Lobelia L.—lobelia, cardinal flower

     *Lobelia erinus L.—edging lobelia, trailing lobelia

     *Lobelia siphilitica L.—great blue lobelia, blue cardinal flower

Nemacladus Nuttall—threadplant, threadstem

Platycodon A. DC.

     *Platycodon grandiflorus (Jacquin) A. DC.—balloon flower

Trachelium L.—throatwort

     *Trachelium rumelianum Hampe—Bulgarian throatwort

Wahlenbergia Schrader ex Roth—royal bluebell

* illustrated

The Remarkable Pollination Mechanism of the Asterales

Many of the Asterales disperse their pollen through a process called plunger pollination. The anthers either stick together in a tube, as seen in the Asteraceae, or cling closely together around the immature style, as in some Campanulaceae. The anthers release their pollen inward, where it is pushed out or picked up by the style as it grows upward through the anthers. Styles of this order may have special projections or other structures that gather the pollen and present it to pollinators. The branches of the style are initially closed. They open and reveal the stigmas only after the style has reached its full height, and pollinators have had an opportunity to gather the pollen.

(↓) The anthers adhere to the young style, (↓) the pollen-covered style elongates, and (↓) the style branches open in balloon flower, *Platycodon grandiflorus*.

Campanulaceae, The Bellflower Family

A clustered bellflower (↑), *Campanula glomerata*, shows the three-branched style of this genus. This harebell flower, bud, and young fruit, *Campanula rotundifolia* (↑), show the small inferior ovary. The corolla of Korean bellflower, *Campanula takesimana*, is white with purple spots inside (↑).

The campanula family has four subfamilies, two of which hold the bellflowers and the lobelias shown here. Although their flowers look different, they have much in common. The ovaries are inferior or half-inferior. The stamens cluster around the immature style. The style grows up through the anthers and pushes or brushes out the pollen as it rises. The stigmas open later, after the style is at its full length. Bellflowers, such as genus *Campanula*, have characteristic bell-shaped, fused corollas with five pointed lobes, and are usually blue, purple, or white. Lobelias have a wider range of color, including red, magenta, and orange. Lobelia flowers are bilaterally symmetrical. They typically twist upside down in the course of their development. Lobelia anthers are usually fused together and form a tube around the style.

Balloon flower (↑), *Platycodon grandiflorus,* is named for its balloon-like buds.

Lobelia erinus (↑) is an annual that is used as a bedding and container plant. Great blue lobelia (↑), *Lobelia siphilitica*, is native to the eastern United States.

Bulgarian throatwort (←), *Trachelium rumelianum,* is an alpine plant with clusters of small flowers.

261

Eudicots

Euasterids II

Dipsacales Dumortier

Adoxaceae Trautvetter   5 genera/200 species

Genera include:

Adoxa L.—moschatel

Sambucus L.

    *Sambucus nigra L.—black elderberry

    **Sambucus racemosa* L.—red elderberry

Viburnum L.

    **Viburnum carlesii* Hemsley—Korean spice viburnum

    **Viburnum plicatum* Thunberg—Japanese snowball

    **Viburnum rhytidophylloides* Suringar—hybrid snowball

* illustrated

Adoxaceae, The Elderberry Family

(↑) Hybrid snowball flowers, *Viburnum rhytidophylloides*, and (↑) its ripening fruits
Korean spice viburnum, *Viburnum carlesii*, is an ornamental known for its fragrance (↑).

This family of shrubs and small trees has opposite leaves that are often toothed, and may be compound or simple. The flowers are small, but masses of them form showy, clustered inflorescences. Individual flowers usually have five petals and stamens, although a few have four-fold parts. The pistil is made of three to five carpels, occasionally fewer. The styles are short if they are there at all, and the stigmas are nestled down in the corolla. The ovary is inferior or half-inferior. The fruits are juicy drupes with one or three to five stones, and are usually red or purple. Many of the shrubs are used in landscaping.

Red elderberry (↑), *Sambucus racemosa*, in flower and fruit (↑), and black elderberry (↑), *S. nigra*, show the compound leaves characteristic of this genus. Black elderberries have been used for centuries to make wine, jelly, and pies. The fruits are usually cooked before they are eaten because they contain antinutritional substances that are destroyed by heating.

Japanese snowball (←), *Viburnum plicatum*, has an adaptation similar to hydrangeas. It has small fertile flowers that are surrounded by larger, showy, sterile flowers.

Eudicots

Euasterids II

Dipsacales

263

Eudicots

Euasterids II

Dipsacales Dumortier

Caprifoliaceae Jussieu   5 genera/220 species

Genera include:

Heptacodium Rehder—Chinese heptacodium, seven son flower

Lonicera L.—honeysuckle

  *Lonicera x *brownii* (Regel) Carrière—Brown's honeysuckle

  *Lonicera x *heckrottii* Rehder—goldflame honeysuckle

  *Lonicera involucrata Banks ex Sprengel—twinberry honeysuckle

Symphoricarpos Duhamel—snowberry, coral berry

  *Symphoricarpos albus (L.) Blake—common snowberry

  *Symphoricarpos occidentalis Hooker—western snowberry

Triosteum L.—wild coffee, horse gentian, feverwort

* illustrated

Ornamental honeysuckles include (↑) goldflame honeysuckle, *Lonicera heck-rottii*, and Brown's honeysuckle, *Lonicera* x *brownii* (↑).

The honeysuckle family is well known for its shrubs, but it also includes vines, small trees, and even herbs. The flowers have a small, persistent calyx and a fused corolla with five lobes. Honeysuckle has a long, thin tube with two lips and protruding stamens and style. The top lip often has four lobes and the bottom lip has one. Snowberry's corolla is a shorter tube that flares open. The styles are longer in Caprifoliaceae than in the related family, Adoxaceae, and can usually be seen extending past the corolla. The ovary is inferior. Fruits of this family are usually berry-like.

Common snowberry (→), *Symphoricarpos albus*, shows the persistent calyx (arrow) on its developing berries. The white berries are poisonous.

The flowers of western snowberry (→), *S. occidentalis*, have a more open bell shape and five similar corolla lobes. Their stamens and pistil (arrow) protrude out of the corolla.

(←) Twinberry, *Lonicera involucrata*, has black fruits backed by showy red bracts. The berries provide food for birds and other wildlife, which act as seed dispersers.

Eudicots

Euasterids II

Dipsacales

Dipsacaceae Jussieu

11 genera/290 species

Genera include:

Cephalaria Schrader ex Roemer & Schultes—giant scabious, yellow cephalaria

Dipsacus L.—teasel

  *Dipsacus fullonum L.—Fuller's teasel

Knautia L.—knautia, blue buttons

Scabiosa L.—scabiosa, pincushion flower

  *Scabiosa stellata L.—starflower pincushion

* illustrated

Dipsacaceae, The Teasel Family

The heads of the Dipsacaceae have involucres and look much like the Asteraceae, but the structure of the flowers shows that this is a different family. For example, flowers of the teasel family usually have four distinct stamens, although two to three occur in a few. There is only one type of flower in the heads, although in *Scabiosa* and related genera, the outer flowers of the head may have much larger corollas than the inner ones (→). The corollas have four or occasionally five fused petals. The style is unbranched and typically has a single knob-like stigma. The flowers have bracts or an epicalyx at their base. The calyx itself is often reduced to bristles, like in Asteraceae. It can also be cup-like.

(→) The prominent styles with their pinhead-like stigmas of *Scabiosa* gives these flowers the common name of pincushion flower. All the flowers mature at about the same time. Fruits of *Scabiosa stellata* (↘) have bristles that are derived from the calyx as well as a papery epicalyx.

Weavers brought teasel (↓), *Dipsacus fullonum*, from Europe to North America. They used its heads for combing the nap of cloth. It has become naturalized and grows in most of the United States. The dry heads are sometimes used in flower arrangements.

Teasel has ovoid heads of small lavender flowers. The bracts of the involucre are nearly as long as the head. They ring the base of the inflorescence and curve upward around it. Each flower has its own pair of small, spiny bracts that form an epicalyx at its base. These bracts become woody and remain after the fruits are mature, (↘) forming the spiny head.

Appendix A. Expanded outline of flowering plant families

This outline includes families that are illustrated in the main section of this book along with additional families that are not illustrated. For the latter, representative genera and common names are listed. This listing is meant to give a larger view of flowering plant families. It concentrates on native plants of temperate North America, as well as horticultural and agricultural plants that are grown in North America, but are native to other areas. It also includes plants used as food or medicine, but grown outside North America, and plants that are significant in the flowering plant tree of life. The groups and order of this listing are based on the 2003 report from the Angiosperm Phylogeny Group, with additional information from the APG website (see reference section). The common names and genera were added from various resources—see the listing of additional resources in the Selected References section.

Basal Angiosperms

Basal lineages
Amborellales
        Amborellaceae (*Amborella*)
Nymphaeales
        Cabombaceae (fanwort – *Cabomba*)
        Nymphaeaceae, the waterlily family
Austrobaileyales
        Illiciaceae (*Illicium* – star anise, anise tree)
        Schisandraceae (*Kudsura* – Japanese kudsura; *Schisandra* – wild sarsaparilla, star vine)

Magnoliids
Laurales
        Calycanthaceae (*Calycanthus* – Carolina allspice, California spice bush; *Chimonathus* – wintersweet)
        Lauraceae, the laurel family
Magnoliales
        Annonaceae (*Annona* – cherimoya, sour sop, sweet sop, pond-apple; *Asimina* – paw-paw; *Deeringothamnus* – Rugel's pawpaw)
        Magnoliaceae, the magnolia family
        Myristicaceae (*Myristica* – nutmeg tree)
Piperales
        Aristolochiaceae, the dutchman's-pipe family
        Piperaceae, the pepper family
        Saururaceae, the lizard's-tail family

Unplaced taxa
Ceratophyllales – uncertain position
        Ceratophyllaceae, sole family (*Ceratophyllum* – water hornwort)
Chloranthales – uncertain position
        Chloranthaceae, sole family (*Chloranthus*)

Monocots

Basal Monocots
Acorales
        Acoraceae, the sweet flag family
Alismatales
        Alismataceae, the water plantain family
        Araceae, the aroids or arum family
        Hydrocharitaceae (*Elodea* – waterweed; *Hydrocharis*; *Limnobium* – frog's-bit; *Vallisneria* – tape grass)
        Potamogetonaceae (*Potamogeton* – pondweeds)
        Zosteraceae (*Phyllospadix* – surf-grass; *Zostera* – eelgrass)

Lilioid or Petaloid Monocots

Dioscoreales
        Dioscoreaceae (*Dioscorea* – tropical yams, air potato; *Tacca* – bat-flower)
Pandanales
        Cyclanthaceae (*Carludovica* – Panama hat plant)
        Pandanaceae (*Pandanus* – screw pine, hala, pandanus)
Asparagales
        Agapanthaceae (native and horticultural species, *Agapanthus*)
        Agavaceae, the agave family
        Alliaceae, the onion family
        Amaryllidaceae, the amaryllis or daffodil family
        Asparagaceae, the asparagus family
        Asphodelaceae (*Aloe*; *Asphodelus* – asphodel; *Bulbine*; *Haworthia*; *Kniphofia* – red-hot poker)
        Hemerocallidaceae, the daylily family
        Hyacinthaceae, the hyacinth family
        Iridaceae, the iris family
        Orchidaceae, the orchid family
        Ruscaceae, the butcher's broom family
        Themidaceae (*Brodiaea*; *Dichelostemma* – firecracker flower)
Liliales
        Alstroemeriaceae, the alstroemeria family
        Colchicaceae (*Colchicum* – autumn crocus; *Gloriosa* – gloriosa lily; *Littonia* –climbing lily; *Uvularia* – bellwort)
        Liliaceae, the lily family
        Melanthiaceae (*Helonias* – swamp pink; *Melanthium* – bunchflower; *Trillium*; *Veratrum* – white hellebore, corn husk lily; *Zigadenus* – wand lily, death camas)
        Smilacaceae (*Smilax* – catbrier, greenbrier, bull brier, sarsaparilla, carrion flower)

Commelinid Monocots

Arecales
        Arecaceae or Palmae, the palm family
Commelinales
        Commelinaceae, the spiderwort family
        Haemodoraceae (*Anigozanthos* – kangaroo paw)
        Pontederiaceae, the pickerel weed or water hyacinth family
Poales
        Bromeliaceae, the bromeliad family
        Cyperaceae, the sedge family
        Juncaceae (*Juncus* – rushes; *Luzula* – wood rushes)
        Poaceae or Gramineae, the grass family
        Sparganiaceae (*Sparganium* – bur-reeds, temperate and Arctic aquatics)
        Typhaceae, the cattail family
        Xyridaceae (*Xyris* – yellow-eyed grass)
Zingiberales
        Cannaceae, the canna family
        Costaceae (*Costus* – spiral flag, ginger lily)
        Heliconiaceae, the heliconia family
        Marantaceae, the prayer plant family
        Musaceae, the banana family
        Strelitziaceae, the bird-of-paradise family
        Zingiberaceae, the ginger family

<u>Eudicots</u> (also called tricolpates, from the three germination openings of the pollen)

Basal Eudicots

Ranunculales
      Berberidaceae, the barberry family
      Menispermaceae (*Anamirta* – fish-berry; *Chondrodendron* – curare; *Cocculus* –Carolina
            moonseed, coralbeads; *Menispermum* – moonseed)
      Papaveraceae, the poppy family and bleeding heart family
      Ranunculaceae, the buttercup family
Buxales
      Buxaceae (*Buxus* – boxwood; *Pachysandra* – Japanese spurge)
Proteales
      Nelumbonaceae, the lotus family
      Platanaceae, the sycamore family
      Proteaceae (*Protea, Banksia, Macadamia, Grevillea*, and many others, mainly Australian and
            South African)

Core Eudicots

Gunnerales
      Gunneraceae (*Gunnera*)
Caryophyllales
      Aizoaceae, the ice plant family
      Amaranthaceae, the amaranth and goosefoot family. This includes the traditional goosefoot
            family, Chenopodiaceae, which is no longer recognized as a separate family.
      Cactaceae, the cactus family
      Caryophyllaceae, the pink family
      Droseraceae, the Venus flytrap family
      Nepenthaceae (*Nepenthes* – Asian pitcher plant)
      Nyctaginaceae, the four o'clock family
      Phytolaccaceae (*Phytolacca* – pokeweed; *Rivinia* – rouge plant)
      Plumbaginaceae, the leadwort family
      Polygonaceae, the buckwheat family
      Portulacaceae, the purslane family
      Sarcobataceae (*Sarcobatus* – greasewood)
      Simmondsiaceae (*Simmondsia* – jojoba)
      Tamaricaceae (*Tamarix* – salt cedar)
Santalales
      Loranthaceae (tropical mistletoes, large-flowered mistletoes)
      Santalaceae, the mistletoe and sandalwood family
Saxifragales
      Altingiaceae (*Liquidambar* – sweet gum)
      Crassulaceae, the stonecrop family
      Grossulariaceae, the gooseberry family
      Haloragaceae (*Myriophyllum* – water-milfoil)
      Hamamelidaceae (*Fothergilla*; *Hamamelis* – witch hazel; *Loropetalum* – Chinese witch hazel;
            Parrotia)
      Iteaceae (*Itea* – sweetspire)
      Paeoniaceae, the peony family
      Saxifragaceae, the saxifrage family

Rosids

Vitales
> Vitaceae, the grape family

Geraniales
> Geraniaceae, the geranium family

Myrtales
> Lythraceae, the loosestrife family
> Melastomataceae (*Rhexia* – meadow beauty; most genera are tropical)
> Myrtaceae, the myrtle family
> Onagraceae, the evening primrose family

Eurosids I

Zygophyllales
> Krameriaceae (*Krameria* – rhatany)
> Zygophyllaceae (*Guaiacum* – lignum vitae; *Larrea* – creosote bush; *Tribulus* –puncture vine)

Celastrales
> Celastraceae (*Celastrus* – bittersweet; *Euonymus*; *Paxistima*)

Cucurbitales
> Begoniaceae, the begonia family
> Cucurbitaceae, the squash family

Fabales
> Fabaceae or Leguminosae – the legume family
>> Caesalpinioideae, the caesalpinia subfamily
>> Cercideae, the redbud and bahinia tribe
>> Faboideae, the bean subfamily
>> Mimosoideae, the mimosa subfamily
> Polygalaceae (*Polygala* – milkwort, snakeroot)

Fagales
> Betulaceae, the birch and alder family
> Casuarinaceae (*Casuarina* – casuarina, ironwood, she-oak, beefwood, Australian pine)
> Fagaceae, the oak family
> Juglandaceae, the walnut family
> Myricaceae (*Myrica* – bayberry, candleberry, wax myrtle)

Rosales
> Cannabaceae (*Cannabis* – hemp, marihuana; *Celtis* – hackberry tree; *Humulus* –hops)
> Elaeagnaceae (*Elaeagnus* – Russian olive, silverberry; *Shepherdia* – buffalo berry)
> Moraceae, the fig and mulberry family
> Rhamnaceae, the buckthorn family
> Rosaceae, the rose family
> Ulmaceae, the elm family
> Urticaceae, the nettle family

Malpighiales
> Euphorbiaceae, the spurge family
> Hypericaceae (*Hypericum* – St. John's wort, Klamath weed)
> Linaceae, the flax family
> Malpighiaceae (*Malpighia* – Barbados cherry)
> Passifloraceae, the passionflower family
> Salicaceae, the willow family
> Violaceae, the violet and pansy family

Oxalidales
> Cephalotaceae (*Cephalotus* – Australian pitcher plant)
> Oxalidaceae, the oxalis family

Eurosids II

Brassicales

    Brassicaceae or Cruciferae, the mustard family

    Capparaceae (*Capparis* – capers)

    Caricaceae (*Carica* – papaya)

    Cleomaceae (*Cleome* – spider flower, bee plant)

    Limnanthaceae (*Limnanthes* – meadow foam, poached eggs)

    Resedaceae (*Reseda* – mignonette)

    Tropaeolaceae (*Tropaeolum* – nasturtiums, canary creeper)

Malvales

    Cistaceae (*Cistus* – rockrose; *Helianthemum* – sunrose; *Hudsonia* – beach heather)

    Malvaceae, the mallow family

        Bombacoideae, the bombax subfamily (*Adansonia* – baobab; *Bombax* – silk tree; *Ceiba* – kapok tree; *Ochroma* – balsawood)

        Byttnerioideae, the cacao subfamily

        Malvoideae, the mallow subfamily

        Sterculioideae (*Bradychiton*; *Cola* – kola nut; *Sterculia* – flame tree)

        Tilioideae, the tilia subfamily

    Thymelaeaceae (*Daphne*)

Sapindales

    Anacardiaceae, the sumac family

    Burseraceae (*Bursera* – gumbo-limbo tree, elephant tree)

    Meliaceae (*Melia* – chinaberry tree; *Swietenia* – mahogany)

    Rutaceae, the citrus family

    Sapindaceae, the soapberry, maple, and horse chestnut family

        Sapindoideae, the soapberry subfamily

        Hippocastanoideae, the horse chestnut and maple subfamily

    Simaroubaceae (*Ailanthus* – tree-of-heaven; *Simarouba* – paradise tree)

Asterids

Cornales

    Cornaceae, the dogwood family

    Hydrangeaceae, the hydrangea family

    Loasaceae, the blazingstar or stickleaf family

    Nyssaceae (*Davidia* – handkerchief or dove tree; *Nyssa* – tupelo tree)

Ericales

    Actinidiaceae (*Actinidia* – kiwi fruit, Chinese gooseberry)

    Balsaminaceae, the impatiens family

    Clethraceae (*Clethra* – lily-of-the-valley tree, summersweet, sweet pepperbush)

    Cyrillaceae (*Cliftonia* – buckwheat tree; *Cyrilla* – leatherwood, titi)

    Diapensiaceae (*Diapensia* – pincushion plant; *Galax* – beetleweed; *Pyxidanthera* –pyxie; *Shortia* – Oconee bells, fringe bells)

    Ebenaceae (*Diospyros* – ebony wood, persimmon)

    Ericaceae, the heath and heather family

    Fouquieriaceae (*Fouquieria* – ocotillo)

    Lecythidaceae (*Bertholletia* – Brazil nuts; *Couroupita* – cannonball tree)

    Myrsinaceae (*Anagallis* – scarlet and blue pimpernels; *Ardisia* – marlberry; *Cyclamen*; *Lysimachia* – yellow loosestrife; *Trientalis* – starflower)

    Polemoniaceae, the phlox family

    Primulaceae, the primrose family

    Sapotaceae (*Manilkara* – sapodilla or chicle; *Sideroxylon* – bumelia, ironwood)

    Sarraceniaceae, the pitcher plant family

    Styracaceae (*Halesia* – silver bell trees, snowdrop tree; *Styrax* – snowbell)

    Symplocaceae (*Symplocos* – sapphireberry, sweetleaf)

    Theaceae (*Camellia* – tea, camellia; *Franklinia* – Franklin tree; *Gordonia* – loblolly-bay)

Euasterids I

Boraginaceae, the borage and waterleaf family. It is not yet placed in an order. This contains the traditional waterleaf family, Hydrophyllaceae, which is no longer recognized as a separate family.

Garryales

Garryaceae (*Arcuba* – Japanese arcuba; *Garrya* – silktassel)

Gentianales

Apocynaceae, the dogbane and milkweed family. This includes the traditional milkweed family, Asclepiaceae, which is no longer recognized as a separate family.

Gelsemiaceae (*Gelsemium* – yellow jessamine, swamp jessamine)

Gentianaceae, the gentian family

Loganiaceae (*Spigelia* – Indian pink, pink root)

Rubiaceae, the madder family

Lamiales

Acanthaceae, the acanthus family

Bignoniaceae, the trumpet vine family

Calceolariaceae (*Calceolaria* – pocket book flower)

Gesneriaceae (*Achimenes* – hot water plant; *Aeschynanthus* – lipstick plant, zebra vine; *Columnea*; *Episcia* – flame violet; *Gesneria* – firecracker; *Saintpaulia* –African violet; *Sinningia* – gloxinia; *Streptocarpus* – cape primrose)

Lamiaceae or Labiatae, the mint family

Lentibulariaceae (*Pinguicula* – butterwort; *Utricularia* – bladderwort)

Martyniaceae (*Martynia, Proboscidea* – unicorn plant, devil's claw)

Oleaceae, the olive family

Orobanchaceae, the broomrape and paintbrush family

Paulowniaceae (*Paulownia* – empress tree)

Pedaliaceae (*Sesamum* – sesame)

Phrymaceae (*Mazus*; *Mimulus* – monkey flower; *Phryma* – lopseed)

Plantaginaceae, the plantain and snapdragon family

Scrophulariaceae, the figwort family

Verbenaceae, the verbena family

Solanales

Convolvulaceae, the morning glory family

Solanaceae, the nightshade family

Euasterids II

Aquifoliales

Aquifoliaceae, the holly family

Apiales

Apiaceae or Umbelliferae, the carrot and parsley family

Araliaceae – (*Aralia*; *Fatsia*; *Hedera* – English ivy; *Panax* – ginseng; *Schefflera* –umbrella plant; *Tetrapanax* – Chinese rice paper plant)

Pittosporaceae (*Pittosporum*)

Asterales

Asteraceae or Compositae, the sunflower family

Carduoideae, the thistle subfamily

Cardueae, the thistle tribe

Cichorioideae, the lettuce subfamily

Arctoteae (*Arctotis* – African daisy)

Cichorieae (Lactuceae), the lettuce tribe

Vernonieae (*Vernonia* – ironweed, *Stokesia*—Stoke's aster)

Asteroideae, the aster subfamily
    Anthemideae, the anthemis tribe
    Astereae, the aster tribe
    Bahieae (*Bahia* – bahia; *Hymenopappus* – white ragweed, old plainsman;
        Palafoxia – palafox)
    Calenduleae (*Calendula* – calendula, pot marigold; *Dimorphotheca* – cape marigold,
        African daisy; *Osteospermum* – African daisy)
    Coreopsideae, the coreopsis tribe
    Eupatorieae, the gayfeather tribe
    Gnaphalieae, the everlastings tribe
    Helenieae, the sneezeweed tribe
    Heliantheae, the sunflower tribe
    Inuleae (*Inula* – elecampane)
    Madieae (*Argyroxiphium* – Hawaiian silversword; *Arnica* – arnica; *Layia* – tidytips;
        Madia – coastal tarweed)
    Senecioneae, the senecio tribe
    Tageteae (*Flaveria* – yellow tops, marshweed; *Pectis* – chinchweed;
        Tagetes – marigold; *Thymophylla* – dogweed, prickly leaf)
Campanulaceae, the bellflower family
Goodeniaceae (*Scaevola* – fanflower)
Menyanthaceae (*Menyanthes* – buckbean, bogbean; *Nymphoides* – floating heart)
Dipsacales
    Adoxaceae, the elderberry family
    Caprifoliaceae, the honeysuckle family
    Diervillaceae (*Diervilla* – bush honeysuckle; *Weigela*)
    Dipsacaceae, the teasel family
    Linnaeaceae (*Abelia*; *Kolkwitzia* – beauty bush; *Linnaea* – twinflower)
    Valerianaceae (*Centranthus* – red valerian; *Patrinia*; *Valeriana* – valerian)

Appendix B. Genus to common name, family, and order listing

Genus	Common name	Family	Order
Abelia	abelia	Linnaeaceae	Dipsacales
Abelmoschus	okra	Malvaceae	Malvales
Abronia	sand verbena	Nyctaginaceae	Caryophyllales
Abutilon	flowering maple, velvetleaf, Chinese bellflower, Chinese jute	Malvaceae	Malvales
Acacia	acacia, wattle tree	Fabaceae or Leguminosae	Fabales
Acalypha	chenille plant	Euphorbiaceae	Malpighiales
Acatholimon	prickly thrift	Plumbaginaceae	Caryophyllales
Acanthus	bear's-breech, oyster plant	Acanthaceae	Lamiales
Acca	pineapple guava, feijoa	Myrtaceae	Myrtales
Acer	maple	Sapindaceae	Sapindales
Achillea	yarrow, milfoil	Asteraceae or Compositae	Asterales
Achimenes	hot water plant	Gesneriaceae	Lamiales
Achlys	vanilla leaf	Berberidaceae	Ranunculales
Acleisanthes	angel trumpets	Nyctaginaceae	Caryophyllales
Aconitum	monkshood	Ranunculaceae	Ranunculales
Acorus	sweet flag	Acoraceae	Acorales
Actaea	baneberry	Ranunculaceae	Ranunculales
Actinidia	kiwi fruit, Chinese gooseberry	Actinidiaceae	Ericales
Adenophora	ladybells	Campanulaceae	Asterales
Adenostoma	chamise, greasewood	Rosaceae	Rosales
Adoxa	moschatel, muskroot	Adoxaceae	Dipsacales
Aechmea	bromeliad	Bromeliaceae	Poales
Aegopodium	bishop's weed, goutweed	Apiaceae or Umbelliferae	Apiales
Aeonium	pinwheel, Canary Island rose	Crassulaceae	Saxifragales
Aeschynanthus	lipstick plant	Gesneriaceae	Lamiales
Aesculus	horse chestnut, buckeye	Sapindaceae	Sapindales
Agalinis	gerardia, false foxglove	Orobanchaceae	Lamiales
Agapanthus	agapanthus	Agapanthaceae	Asparagales
Agastache	nettleleaf horsemint, giant hyssop, hummingbird mint, sunset hyssop	Lamiaceae or Labiatae	Lamiales
Agave	century plant	Agavaceae	Asparagales
Ageratina	snakeroot	Asteraceae or Compositae	Asterales
Ageratum	floss flower	Asteraceae or Compositae	Asterales
Aglaonema	Chinese evergreen	Araceae	Alismatales
Agrimonia	agrimony	Rosaceae	Rosales
Agrostemma	corn cockle	Caryophyllaceae	Caryophyllales
Ailanthus	tree-of-heaven	Simaroubaceae	Sapindales
Ajuga	ajuga, carpet bugle	Lamiaceae or Labiatae	Lamiales
Albizia	silk tree	Fabaceae or Leguminosae	Fabales
Alcantarea	imperial bromeliad	Bromeliaceae	Poales
Alcea	hollyhock	Malvaceae	Malvales
Alchemilla	lady's-mantle	Rosaceae	Rosales
Aldrovanda	waterwheel plant	Droseraceae	Caryophyllales
Aletes	mountain caraway	Apiaceae or Umbelliferae	Apiales

Alisma	water plantain	Alismataceae	Alismatales
Allionia	trailing four o'clock	Nyctaginaceae	Caryophyllales
Allium	onion, leek, garlic, and native and ornamental alliums	Alliaceae	Asparagales
Alnus	alder	Betulaceae	Fagales
Alocasia	elephant's ear	Araceae	Alismatales
Aloe	aloe	Asphodelaceae	Asparagales
Aloysia	bee brush, white brush	Verbenaceae	Lamiales
Alpinia	shell ginger, galangal	Zingiberaceae	Zingiberales
Alstroemeria	alstroemeria, Chilean lily, Peruvian lily	Alstroemeriaceae	Liliales
Alternanthera	khakiweed, alligator weed	Amaranthaceae	Caryophyllales
Althaea	marsh mallow	Malvaceae	Malvales
Alyogyne	blue hibiscus	Malvaceae	Malvales
Alyssum	perennial alyssum	Brassicaceae or Cruciferae	Brassicales
Amaranthus	pigweed, careless weed	Amaranthaceae	Caryophyllales
Amaryllis	naked lady, Cape belladonna	Amaryllidaceae	Asparagales
Amborella	—	Amborellaceae	Amborellales
Ambrosia	ragweed	Asteraceae or Compositae	Asterales
Amelanchier	service berry, saskatoon	Rosaceae	Rosales
Amorpha	leadplant	Fabaceae or Leguminosae	Fabales
Ampelopsis	blueberry climber, porcelain berry	Vitaceae	Vitales
Amsinckia	fiddleneck	Boraginaceae	not assigned
Amsonia	bluestar	Apocynaceae	Gentianales
Amyris	torchwood	Rutaceae	Sapindales
Anacardium	cashew	Anacardiaceae	Sapindales
Anacyclus	Atlas daisy	Asteraceae or Compositae	Asterales
Anagallis	pimpernel	Myrsinaceae	Ericales
Ananas	pineapple	Bromeliaceae	Poales
Anaphalis	pearly everlasting	Asteraceae or Compositae	Asterales
Anchusa	summer forget-me-not	Boraginaceae	not assigned
Andromeda	bog rosemary	Ericaceae	Ericales
Andropogon	bluestem grass	Poaceae	Poales
Androsace	rock jasmine	Primulaceae	Ericales
Anemone	windflowers, hepatica	Ranunculaceae	Ranunculales
Anemopsis	yerba mansa	Saururaceae	Piperales
Anethum	dill	Apiaceae or Umbelliferae	Apiales
Angelica	angelica	Apiaceae or Umbelliferae	Apiales
Annona	cherimoya, sour sop, sweet sop, pond-apple	Annonaceae	Magnoliales
Antennaria	pussytoes	Asteraceae or Compositae	Asterales
Anthemis	golden marguerite	Asteraceae or Compositae	Asterales
Anthriscus	chervil	Apiaceae or Umbelliferae	Apiales
Anthurium	anthurium	Araceae	Alismatales
Antigonon	coral vine	Polygonaceae	Caryophyllales
Antirrhinum	snapdragons	Plantaginaceae	Lamiales
Aphelandra	zebra plant	Acanthaceae	Lamiales
Apium	celery	Apiaceae or Umbelliferae	Apiales
Apocynum	dogbanes	Apocynaceae	Gentianales

Apodanthera	melon loco	Cucurbitaceae	Cucurbitales
Aptenia	heartleaf ice plant	Aizoaceae	Caryophyllales
Aquilegia	columbines	Ranunculaceae	Ranunculales
Arabis	rockcress	Brassicaceae or Cruciferae	Brassicales
Arachis	peanuts	Fabaceae or Leguminosae	Fabales
Aralia	Hercules club	Araliaceae	Apiales
Arbutus	madrone, strawberry tree	Ericaceae	Ericales
Arceuthobium	western dwarf mistletoe	Santalaceae	Santalales
Arctium	burdock	Asteraceae or Compositae	Asterales
Arctostaphylos	kinnikinnick	Ericaceae	Ericales
Ardisia	marlberry	Myrsinaceae	Ericales
Arenaria	sandwort	Caryophyllaceae	Caryophyllales
Argemone	prickly poppy	Papaveraceae	Ranunculales
Argyroxiphium	Hawaiian silversword	Asteraceae or Compositae	Asterales
Arisaema	Jack-in-the-pulpit, green dragon	Araceae	Alismatales
Aristolochia	Dutchman's pipe, pipe vine, birthwort	Aristolochiaceae	Piperales
Armeria	sea pink, thrift	Plumbaginaceae	Caryophyllales
Armoracia	horseradish	Brassicaceae or Cruciferae	Brassicales
Arnica	arnica	Asteraceae or Compositae	Asterales
Arnoglossum	Indian plantain	Asteraceae or Compositae	Asterales
Aronia	chokeberry	Rosaceae	Rosales
Artemisia	mugwort, wormwood, tarragon, southernwood, dusty miller, big sagebrush	Asteraceae or Compositae	Asterales
Artocarpus	breadfruit	Moraceae	Rosales
Aruncus	goatsbeard	Rosaceae	Rosales
Asarina	asarina, twining snapdragon	Plantaginaceae	Lamiales
Asarum	wild ginger, little brown jugs	Aristolochiaceae	Piperales
Asclepias	milkweed, butterfly weed	Apocynaceae	Gentianales
Asimina	paw-paw	Annonaceae	Magnoliales
Asparagus	asparagus	Asparagaceae	Asparagales
Asphodelus	asphodel	Asphodelaceae	Asparagales
Aster	asters	Asteraceae or Compositae	Asterales
Astilbe	meadow sweet	Saxifragaceae	Saxifragales
Astragalus	locoweed	Fabaceae or Leguminosae	Fabales
Astrantia	masterwort	Apiaceae or Umbelliferae	Apiales
Atriplex	salt bush, orache	Amaranthaceae	Caryophyllales
Atropa	belladonna	Solanaceae	Solanales
Aubrieta	aubrieta	Brassicaceae or Cruciferae	Brassicales
Aucuba	Japanese aucuba	Garryaceae	Garryales
Aurinia	basket-of-gold	Brassicaceae or Cruciferae	Brassicales
Avena	oats	Poaceae or Gramineae	Poales
Averrhoa	star fruit, carambola	Oxalidaceae	Oxalidales
Avicennia	black mangrove	Acanthaceae	Lamiales
Bacopa	bacopa	Plantaginaceae	Lamiales
Baileya	desert marigold	Asteraceae or Compositae	Asterales
Balduina	honeycomb head	Asteraceae or Compositae	Asterales
Balsamorhiza	balsamroot	Asteraceae or Compositae	Asterales
Bambusa	bamboo	Poaceae or Gramineae	Poales

278

Baptisia	wild indigo	Fabaceae or Leguminosae	Fabales
Bauhinia	orchid tree	Fabaceae or Leguminosae	Fabales
Begonia	begonia	Begoniaceae	Cucurbitales
Belamcanda	blackberry lily	Iridaceae	Asparagales
Bellis	English daisy	Asteraceae or Compositae	Asterales
Berberis	barberry	Berberidaceae	Ranunculales
Bergenia	bergenia	Saxifragaceae	Saxifragales
Berlandiera	chocolate daisy, green eyes	Asteraceae or Compositae	Asterales
Bertholletia	Brazil nut	Lecythidaceae	Ericales
Berula	water parsnip	Apiaceae or Umbelliferae	Apiales
Besseya	coral drops, kittentails	Plantaginaceae	Lamiales
Beta	beet, chard	Amaranthaceae	Caryophyllales
Betula	birch	Betulaceae	Fagales
Bidens	beggarticks	Asteraceae or Compositae	Asterales
Bignonia	cross vine	Bignoniaceae	Lamiales
Billbergia	Queen's tears, bromeliad	Bromeliaceae	Poales
Bistorta	bistort	Polygonaceae	Caryophyllales
Bletilla	Chinese ground orchid	Orchidaceae	Asparagales
Boehmeria	ramie, false nettle	Urticaceae	Rosales
Boerhavia	spiderlings	Nyctaginaceae	Caryophyllales
Boisduvalia	spike primrose	Onagraceae	Myrtales
Boltonia	boltonia, false aster	Asteraceae or Compositae	Asterales
Borago	borage	Boraginaceae	not assigned
Bougainvillea	bougainvillea	Nyctaginaceae	Caryophyllales
Bouteloua	grama grass	Poaceae or Gramineae	Poales
Bouvardia	bouvardia	Rubiaceae	Gentianales
Boykinia	boykinia	Saxifragaceae	Saxifragales
Brachycome	Swan River daisy	Asteraceae or Compositae	Asterales
Brassica	cabbage, broccoli, cauliflower, kale, kohlrabi, canola, turnip, Brussels sprouts, mustard weed	Brassicaceae or Cruciferae	Brassicales
Brickellia	brickelbush	Asteraceae or Compositae	Asterales
Brimeura	brimeura	Hyacinthaceae	Asparagales
Briza	quaking grass	Poaceae or Gramineae	Poales
Brodiaea	brodiaea	Themidaceae	Asparagales
Bromus	brome grass	Poaceae or Gramineae	Poales
Broussonetia	paper mulberry	Moraceae	Rosales
Browallia	amethyst flower	Solanaceae	Solanales
Brugmansia	angel's trumpet	Solanaceae	Solanales
Brunfelsia	yesterday-today-and-tomorrow	Solanaceae	Solanales
Brunnera	brunnera, Siberian bugloss	Boraginaceae	not assigned
Buchloe	buffalo grass	Poaceae or Gramineae	Poales
Buddleja	butterfly bush	Scrophulariaceae	Lamiales
Bumelia	bumelia, saffron-plum	Sapotaceae	Ericales
Bupleurum	thoroughwax	Apiaceae or Umbelliferae	Apiales
Bursera	gumbo-limbo tree, elephant tree	Burseraceae	Sapindales
Buxus	boxwood	Buxaceae	Buxales
Cabomba	fanwort	Cabombaceae	Nymphaeales
Caesalpinia	red bird-of-paradise, holdback, knickers, Barbados pride	Fabaceae	Fabales

Caladium	caladium	Araceae	Alismatales
Calamus	rattan palm	Arecaceae or Palmae	Arecales
Calandrinia	red maids, rock purslane	Portulacaceae	Caryophyllales
Calathea	peacock plant	Marantaceae	Zingiberales
Calceolaria	pocketbook flower	Calceolariaceae	Lamiales
Calendula	pot marigold	Asteraceae or Compositae	Asterales
Calibrachoa	million bells, miniature petunia	Solanaceae	Solanales
Calliandra	powder puff	Fabaceae or Leguminosae	Fabales
Callicarpa	beauty berry	Lamiaceae or Labiatae	Lamiales
Callirhoe	wine cups, poppy mallow	Malvaceae	Malvales
Callisia	inch plant, basket plant	Commelinaceae	Commelinales
Callistemon	bottlebrush	Myrtaceae	Myrtales
Callistephus	China aster	Asteraceae or Compositae	Asterales
Callitriche	water-starwort	Plantaginaceae	Lamiales
Calluna	heather	Ericaceae	Ericales
Calochortus	mariposa or sago lily	Liliaceae	Liliales
Caltha	marsh marigold	Ranunculaceae	Ranunculales
Calycanthus	Carolina allspice, California spice bush	Calycanthaceae	Laurales
Calypso	fairy slipper	Orchidaceae	Asparagales
Calystegia	hedge bindweed	Convolvulaceae	Solanales
Camassia	camass	Agavaceae	Asparagales
Camellia	camellia, tea	Theaceae	Ericales
Camissonia	evening primrose, sun cups	Onagraceae	Myrtales
Campanula	bellflower, Canterbury bells, harebell, bluebell of Scotland	Campanulaceae	Asterales
Campsis	trumpet creeper, trumpet vine	Bignoniaceae	Lamiales
Canna	canna	Cannaceae	Zingiberales
Cannabis	hemp, marihuana	Cannabaceae	Rosales
Capparis	capers	Capparaceae	Brassicales
Capsella	shepherd's purse	Brassicaceae or Cruciferae	Brassicales
Capsicum	bell pepper, chili pepper, pimento, jalapeno, paprika	Solanaceae	Solanales
Cardamine	milkmaids, bittercress, toothwort, cuckoo flower	Brassicaceae	Brassicales
Carduus	thistle, musk thistle, Italian thistle, bristle thistle	Asteraceae or Compositae	Asterales
Carex	sedge	Cyperaceae	Poales
Carica	papaya	Caricaceae	Brassicales
Carludovica	Panama hat plant	Cyclanthaceae	Pandanales
Carnegiea	saguaro	Cactaceae	Caryophyllales
Carpenteria	tree anemone, bush anemone	Hydrangeaceae	Cornales
Carphephorus	chaffhead, vanilla leaf	Asteraceae or Compositae	Asterales
Carpinus	ironwood, hornbeam	Betulaceae	Fagales
Carpobrotus	ice plant	Aizoaceae	Caryophyllales
Carthamus	safflower, distaff thistle	Asteraceae or Compositae	Asterales
Carum	caraway	Apiaceae or Umbelliferae	Apiales
Carya	pecan, hickory nut	Juglandaceae	Fagales
Caryopteris	blue spirea, bluebeard, blue mist	Lamiaceae or Labiatae	Lamiales
Caryota	fishtail palm	Arecaceae or Palmae	Arecales

Cassia	golden shower tree, pink shower	Fabaceae or Leguminosae	Fabales
Cassiope	white mountain heather	Ericaceae	Ericales
Castanea	chestnut	Fagaceae	Fagales
Castanopsis	chinkapin	Fagaceae	Fagales
Castilleja	Indian paintbrush, paintbrush	Orobanchaceae	Lamiales
Casuarina	she-oak, beefwood	Casuarinaceae	Fagales
Catalpa	catalpa tree	Bignoniaceae	Lamiales
Catananche	cupid's dart	Asteraceae or Compositae	Asterales
Catharanthus	Madagascar periwinkle	Apocynaceae	Gentianales
Cattleya	cattleya orchid	Orchidaceae	Asparagales
Caulophyllum	blue cohosh	Berberidaceae	Ranunculales
Ceanothus	California lilac, New Jersey tea, redroot, buckbrush	Rhamnaceae	Rosales
Celastrus	bittersweet	Celastraceae	Celastrales
Celosia	cockscomb	Amaranthaceae	Caryophyllales
Celtis	hackberry tree	Cannabaceae	Rosales
Centaurea	cornflower, knapweed, starthistle	Asteraceae or Compositae	Asterales
Centaurium	centaury, Rosita	Gentianaceae	Gentianales
Centranthus	red valerian, Jupiter's beard	Valerianaceae	Dipsacales
Cephalanthus	buttonbush, button willow	Rubiaceae	Gentianales
Cephalaria	cephalaria, giant scabious	Dipsacaceae	Dipsacales
Cephalophyllum	ice plant	Aizoaceae	Caryophyllales
Cerastium	chickweed, snow-in-summer	Caryophyllaceae	Caryophyllales
Ceratophyllum	water hornwort	Ceratophyllaceae	Ceratophyllales
Ceratostigma	plumbago	Plumbaginaceae	Caryophyllales
Cercidium	see *Parkinsonia*		
Cercis	redbud	Fabaceae or Leguminosae	Fabales
Cercocarpus	mountain mahogany	Rosaceae	Rosales
Cereus	hedge cactus, queen-of-the-night	Cactaceae	Caryophyllales
Cestrum	night jasmine	Solanaceae	Solanales
Cevallia	stinging stick-leaf	Loasaceae	Cornales
Chamaedaphne	leatherleaf	Ericaceae	Ericales
Chamaemelum	chamomile (ground cover)	Asteraceae or Compositae	Asterales
Chamaesyce	prostrate spurge	Euphorbiaceae	Malpighiales
Chamelaucium	waxflower (florists')	Myrtaceae	Myrtales
Chelidonium	celandine	Papaveraceae	Ranunculales
Chelone	turtlehead	Plantaginaceae	Lamiales
Chenopodium	goosefoot, lamb's quarters, strawberry blite, quinoa	Amaranthaceae	Caryophyllales
Chilopsis	desert willow	Bignoniaceae	Lamiales
Chimaphila	pipsissewa	Ericaceae	Ericales
Chimonanthus	wintersweet	Calycanthaceae	Laurales
Chionanthus	fringe tree	Oleaceae	Lamiales
Chionodoxa	glory-of-the-snow	Hyacinthaceae	Asparagales
Chloranthus	chloranthus	Chloranthaceae	Chloranthales
Chlorogalum	soap plant, amole	Agavaceae	Asparagales
Choisya	Mexican orange	Rutaceae	Sapindales
Chondrilla	hogbite, skeleton weed	Asteraceae or Compositae	Asterales

Chorispora	purple mustard weed	Brassicaceae or Cruciferae	Brassicales
Chrysanthemum	chrysanthemum, painted daisy, marguerite, costmary	Asteraceae or Compositae	Asterales
Chrysolepis	western chinkapin, sierra chinkapin	Fagaceae	Fagales
Chrysopsis	goldenaster	Asteraceae or Compositae	Asterales
Chrysosplenium	golden carpet, golden saxifrage	Saxifragaceae	Saxifragales
Chrysothamnus	see *Ericameria*		
Cichorium	chicory, endive, radicchio	Asteraceae or Compositae	Asterales
Cicuta	water hemlock	Apiaceae or Umbelliferae	Apiales
Cimicifuga	black cohosh	Ranunculaceae	Ranunculales
Cinchona	quinine	Rubiaceae	Gentianales
Cineraria	cineraria (yellow)	Asteraceae or Compositae	Asterales
Cinnamomum	cinnamon, camphor tree	Lauraceae	Laurales
Circaea	enchanter's nightshade	Onagraceae	Myrtales
Cirsium	thistle, Canadian thistle	Asteraceae or Compositae	Asterales
Cissus	grape ivy	Vitaceae	Vitales
Cistus	rockrose	Cistaceae	Malvales
Citrullus	watermelon	Cucurbitaceae	Cucurbitales
Citrus	orange, lemon, lime, grapefruit	Rutaceae	Sapindales
Clarkia	godetia, farewell-to-spring	Onagraceae	Myrtales
Claytonia	spring beauty, miner's lettuce	Portulacaceae	Caryophyllales
Clematis	clematis, virgin's bower, sugar bowls	Ranunculaceae	Ranunculales
Cleome	spider flower, bee plant	Cleomaceae	Brassicales
Clerodendrum	bleeding heart vine, pagoda flower, glorybower	Lamiaceae or Labiatae	Lamiales
Clethra	lily-of-the-valley tree, sweet pepperbush, summersweet	Clethraceae	Ericales
Cliftonia	buckwheat tree	Cyrillaceae	Ericales
Clintonia	bluebead lily	Liliaceae	Liliales
Clivia	clivia, kaffir lily	Amaryllidaceae	Asparagales
Cnidoscolus	spurge nettle	Euphorbiaceae	Malpighiales
Cobaea	cup-and-saucer vine	Polemoniaceae	Ericales
Coccoloba	sea grape	Polygonaceae	Caryophyllales
Cocculus	coralbeads, Carolina moonseed	Menispermaceae	Ranunculales
Cocos	coconut palm	Arecaceae or Palmae	Arecales
Codiaeum	croton (houseplants, tropical)	Euphorbiaceae	Malpighiales
Codonopsis	bonnet bellflower	Campanulaceae	Asterales
Coffea	coffee	Rubiaceae	Gentianales
Colchicum	autumn crocus	Colchicaceae	Liliales
Coleonema	confetti bush	Rutaceae	Sapindales
Collinsia	Chinese houses, blue-eyed Mary	Plantaginaceae	Lamiales
Collomia	alpine collomia	Polemoniaceae	Ericales
Colocasia	taro, elephant's ear	Araceae	Alismatales
Columnea	column flower	Gesneriaceae	Lamiales
Commelina	dayflower	Commelinaceae	Commelinales
Conium	poison hemlock	Apiaceae or Umbelliferae	Apiales
Conoclinium	blue mist flower, thoroughwort	Asteraceae or Compositae	Asterales
Convallaria	lily-of-the-valley	Ruscaceae	Asparagales

Convolvulus	bindweed, bush morning glory, dwarf morning glory	Convolvulaceae	Solanales
Conyza	horseweed	Asteraceae or Compositae	Asterales
Copernicia	carnuba wax palm	Arecaceae or Palmae	Arecales
Coptis	goldenthread	Ranunculaceae	Ranunculales
Corallorrhiza	coralroot	Orchidaceae	Asparagales
Cordylanthus	birdbeak, club-flower	Orobanchaceae	Lamiales
Coreopsis	coreopsis	Asteraceae or Compositae	Asterales
Coriandrum	coriander, cilantro, Chinese parsley	Apiaceae or Umbelliferae	Apiales
Cornus	dogwood, bunchberry, Cornelian cherry	Cornaceae	Cornales
Corydalis	corydalis, golden smoke	Papaveraceae	Ranunculales
Corylus	hazelnut	Betulaceae	Fagales
Cosmos	cosmos	Asteraceae or Compositae	Asterales
Cotinus	smoke tree	Anacardiaceae	Sapindales
Cotoneaster	cotoneaster	Rosaceae	Rosales
Cowania	cliff rose	Rosaceae	Rosales
Crambe	sea kale, giant kale	Brassicaceae or Cruciferae	Brassicales
Craspedia	drumsticks	Asteraceae or Compositae	Asterales
Crassula	jade plant, crassula	Crassulaceae	Saxifragales
Crataegus	hawthorn	Rosaceae	Rosales
Crepis	hawk's beard	Asteraceae or Compositae	Asterales
Crinum	swamp lily	Amaryllidaceae	Asparagales
Crocosmia	crocosmia	Iridaceae	Asparagales
Crocus	crocus	Iridaceae	Asparagales
Croton	croton (native to SW USA)	Euphorbiaceae	Malpighiales
Cryptantha	hidden flower	Boraginaceae	not assigned
Cryptanthus	earth star bromeliad	Bromeliaceae	Poales
Ctenanthe	ctenanthe	Marantaceae	Zingiberales
Cucumis	cantaloupe, cucumber	Cucurbitaceae	Cucurbitales
Cucurbita	squash, pumpkin, gourd, buffalo gourd, marrow	Cucurbitaceae	Cucurbitales
Cuminum	cumin	Apiaceae or Umbelliferae	Apiales
Cuphea	cuphea, cigar plant, Mexican heather, Hawaiian heather	Lythraceae	Myrtales
Curcuma	turmeric, hidden lily	Zingiberaceae	Zingiberales
Cuscuta	dodder	Convolvulaceae	Solanales
Cyclamen	cyclamen	Myrsinaceae	Ericales
Cynara	globe artichoke, cardoon	Asteraceae or Compositae	Asterales
Cynodon	Bermuda grass	Poaceae or Gramineae	Poales
Cynoglossum	hound's tongue, Chinese forget-me-not	Boraginaceae	not assigned
Cyperus	papyrus	Cyperaceae	Poales
Cypripedium	lady's slipper	Orchidaceae	Asparagales
Cyrilla	leatherwood, titi, cyrilla	Cyrillaceae	Ericales
Cytisus	broom	Fabaceae or Leguminosae	Fabales
Dahlia	dahlia	Asteraceae or Compositae	Asterales
Dalea	dalea, prairie clover, indigo bush	Fabaceae or Leguminosae	Fabales
Damsonium	star water plantain	Alismataceae	Alismatales

Daphne	daphne	Thymelaeaceae	Malvales
Darlingtonia	California pitcher plant, cobra lily	Sarraceniaceae	Ericales
Dasiphora	shrubby cinquefoil	Rosaceae	Rosales
Dasylirion	sotol, desert spoon	Ruscaceae	Asparagales
Datura	jimsonweed, sacred datura	Solanaceae	Solanales
Daucus	carrot, Queen Anne's lace	Apiaceae or Umbelliferae	Apiales
Davidia	dove tree, handkerchief tree	Nyssaceae	Cornales
Decodon	swamp loosestrife	Lythraceae	Myrtales
Decumaria	climbing hydrangea, wood vamp	Hydrangeaceae	Cornales
Deeringothamnus	Rugel's paw-paw	Annonaceae	Magnoliales
Delosperma	ice plant	Aizoaceae	Caryophyllales
Delphinium	delphinium, larkspur	Ranunculaceae	Ranunculales
Dendrobium	orchid	Orchidaceae	Asparagales
Dendromecon	tree poppy	Papaveraceae	Ranunculales
Descurainia	flixweed, tansy-mustard	Brassicaceae or Cruciferae	Brassicales
Deutzia	deutzia	Hydrangeaceae	Cornales
Dianella	flax lily	Hemerocallidaceae	Asparagales
Dianthus	carnation, sweet william, pink	Caryophyllaceae	Caryophyllales
Diapensia	diapensia, pincushion pin	Diapensiaceae	Ericales
Diascia	twinspur	Scrophulariaceae	Lamiales
Dicentra	bleeding hearts, Dutchman's breeches, squirrel corn	Papaveraceae	Ranunculales
Dichelostemma	firecracker flower	Themidaceae	Asparagales
Dichondra	dichondra	Convolvulaceae	Solanales
Dictamnus	gas plant	Rutaceae	Sapindales
Dieffenbachia	dieffenbachia	Araceae	Alismatales
Dierama	fairy wand, wandflower	Iridaceae	Asparagales
Diervilla	bush honeysuckle	Diervillaceae	Dipsacales
Digitalis	foxglove	Plantaginaceae	Lamiales
Digitaria	crabgrass	Poaceae or Gramineae	Poales
Dionaea	Venus fly trap	Droseraceae	Caryophyllales
Dioscorea	tropical yam	Dioscoreaceae	Dioscoreales
Diospyros	persimmon, ebony wood	Ebenaceae	Ericales
Diphylleia	umbrella leaf	Berberidaceae	Ranunculales
Dipsacus	teasel	Dipsacaceae	Dipsacales
Dodecatheon	shooting star	Primulaceae	Ericales
Doronicum	leopard's bane	Asteraceae or Compositae	Asterales
Dorotheanthus	Livingstone daisy, ice plant	Aizoaceae	Caryophyllales
Draba	draba	Brassicaceae or Cruciferae	Brassicales
Dracaena	dracaena, dragon's blood tree, lucky bamboo	Ruscaceae	Asparagales
Dracocephalum	dragon's head	Lamiaceae or Labiatae	Lamiales
Drosera	sundew	Droseraceae	Caryophyllales
Dryas	mountain avens	Rosaceae	Rosales
Dudleya	cliff lettuce, dudleya	Crassulaceae	Saxifragales
Duranta	sky flower, golden dewdrop	Verbenaceae	Lamiales
Ecballium	squirting cucumber	Cucurbitaceae	Cucurbitales
Echeveria	hen and chicks	Crassulaceae	Saxifragales
Echinacea	purple coneflower	Asteraceae or Compositae	Asterales

284

Echinocactus	horse crippler, devil's head	Cactaceae	Caryophyllales
Echinocereus	hedgehog cactus, claret cup cactus	Cactaceae	Caryophyllales
Echinocystis	bur cucumber, wild cucumber	Cucurbitaceae	Cucurbitales
Echinodorus	aquarium swordplant	Alismataceae	Alismatales
Echinops	globe thistle	Asteraceae or Compositae	Asterales
Echium	viper's bugloss	Boraginaceae	not assigned
Eichhornia	water hyacinth	Pontederiaceae	Commelinales
Elaeagnus	Russian olive, silverberry	Elaeagnaceae	Rosales
Eleocharis	spike rush, Chinese water chestnut	Cyperaceae	Poales
Elettaria	cardamom	Zingiberaceae	Zingiberales
Elodea	waterweed	Hydrocharitaceae	Alismatales
Elymus	wheatgrass, wild rye	Poaceae or Gramineae	Poales
Emilia	Flora's paintbrush, tasselflower	Asteraceae or Compositae	Asterales
Empetrum	crowberry	Ericaceae	Ericales
Ensete	Abyssian banana	Musaceae	Zingiberales
Epidendrum	greenfly orchid	Orchidaceae	Asparagales
Epifagus	beech-drops	Orobanchaceae	Lamiales
Epigaea	trailing arbutus	Ericaceae	Ericales
Epilobium	fireweed, willow herb	Onagraceae	Myrtales
Epimedium	bishop's hat	Berberidaceae	Ranunculales
Epiphyllum	orchid cactus	Cactaceae	Caryophyllales
Epipremnum	golden pothos	Araceae	Alismatales
Episcia	flame violet	Gesneriaceae	Lamiales
Eranthis	winter aconite	Ranunculaceae	Ranunculales
Erechtites	burnweed	Asteraceae or Compositae	Asterales
Erica	heath	Ericaceae	Ericales
Ericameria	rabbitbush, golden bush	Asteraceae or Compositae	Asterales
Erigeron	fleabanes, showy daisy	Asteraceae or Compositae	Asterales
Eriogonum	sulfur flower	Polygonaceae	Caryophyllales
Eriophorum	cotton grass	Cyperaceae	Poales
Eritrichium	alpine forget-me-not	Boraginaceae	not assigned
Erodium	storksbill, filaree, heronsbill	Geraniaceae	Geraniales
Eryngium	sea holly, coyote thistle	Apiaceae or Umbelliferae	Apiales
Erysimum	wallflower	Brassicaceae or Cruciferae	Brassicales
Erythrina	coral tree	Fabaceae or Leguminosae	Fabales
Erythronium	trout lily, dog-tooth violet, fawn lily, adder's tongue	Liliaceae	Liliales
Eschscholzia	California poppy	Papaveraceae	Ranunculales
Etlingera	torch ginger	Zingiberaceae	Zingiberales
Eucalyptus	gum tree	Myrtaceae	Myrtales
Eucnide	rock nettle	Loasaceae	Cornales
Eucomis	pineapple lily	Hyacinthaceae	Asparagales
Eugenia	Surinam cherry	Myrtaceae	Myrtales
Euonymus	euonymus, burning bush, winter creeper	Celastraceae	Celastrales
Eupatorium	Joe-pye weed, agrimony, boneset, mist flower	Asteraceae or Compositae	Asterales
Euphorbia	poinsettia, spurge, crown-of-thorns, snow-on-the-mountain	Euphorbiaceae	Malpighiales

Euphrasia	eyebright	Orobanchaceae	Lamiales
Eustoma	lisianthus, prairie gentian	Gentianaceae	Gentianales
Exacum	German or Persian violet	Gentianaceae	Gentianales
Exothea	inkwood	Sapindaceae	Sapindales
Fagopyrum	buckwheat	Polygonaceae	Caryophyllales
Fagus	beech	Fagaceae	Fagales
Fallopia	silver lace vine, Japanese knotweed	Polygonaceae	Caryophyllales
Fallugia	Apache plume	Rosaceae	Rosales
Fatsia	Japanese fatsia	Araliaceae	Apiales
Feijoa	see *Acca*		
Felicia	blue marguerite, shrub aster	Asteraceae or Compositae	Asterales
Fendlera	fendlerbush	Hydrangeaceae	Cornales
Fenestraria	babytoes	Aizoaceae	Caryophyllales
Ferocactus	barrel cactus	Cactaceae	Caryophyllales
Festuca	fescue grass	Poaceae or Gramineae	Poales
Ficus	figs, banyan tree	Moraceae	Rosales
Filago	cottonrose	Asteraceae or Compositae	Asterales
Filipendula	queen-of-the-prairie, queen-of-the-meadow	Rosaceae	Rosales
Fittonia	nerve plant	Acanthaceae	Lamiales
Foeniculum	fennel	Apiaceae or Umbelliferae	Apiales
Forestiera	swamp privet, desert olive	Oleaceae	Lamiales
Forsythia	forsythia, golden bells	Oleaceae	Lamiales
Fortunella	kumquat	Rutaceae	Sapindales
Fothergilla	fothergilla	Hamamelidaceae	Saxifragales
Fouquieria	ocotillo, boojum	Fouquieriaceae	Ericales
Fragaria	strawberry	Rosaceae	Rosales
Franklinia	Franklin tree	Theaceae	Ericales
Frasera	monument plant, green gentian	Gentianaceae	Gentianales
Fraxinus	ash tree	Oleaceae	Lamiales
Freesia	freesia	Iridaceae	Asparagales
Fremontodendron	flannelbush	Malvaceae	Malvales
Fritillaria	fritillaria, checker lily	Liliaceae	Liliales
Froelichia	snakecotton, cottonweed	Amaranthaceae	Caryophyllales
Fuchsia	fuchsia	Onagraceae	Myrtales
Fumaria	fumitory	Papaveraceae	Ranunculales
Gaillardia	blanket flower	Asteraceae or Compositae	Asterales
Galanthus	snowdrops	Amaryllidaceae	Asparagales
Galax	galax, beetleweed	Diapensiaceae	Ericales
Galinsoga	quickweed	Asteraceae or Compositae	Asterales
Galium	bedstraw, sweet woodruff, goose grass, cleavers, wild madder	Rubiaceae	Gentianales
Galtonia	summer hyacinth	Hyacinthaceae	Asparagales
Gardenia	gardenia	Rubiaceae	Gentianales
Garrya	silktassel	Garryaceae	Garryales
Gaultheria	wintergreen, salal	Ericaceae	Ericales
Gaura	gaura, beeblossom	Onagraceae	Myrtales
Gelsemium	yellow jessamine, swamp jessamine	Gelsemiaceae	Gentianales

Genista	broom	Fabaceae or Leguminosae	Fabales
Gentiana	gentian	Gentianaceae	Gentianales
Gentianella	dwarf gentian	Gentianaceae	Gentianales
Gentianopsis	fringed gentian	Gentianaceae	Gentianales
Geranium	wild geranium, hearty geranium, cranesbill, herb Robert	Geraniaceae	Geraniales
Gesneria	firecracker	Gesneriaceae	Lamiales
Geum	geum, avens, prairie smoke	Rosaceae	Rosales
Gilia	bird's eyes, Queen Anne's thimbles, globe gilia	Polemoniaceae	Ericales
Gladiolus	glads, gladiolus	Iridaceae	Asparagales
Glandularia	mock vervain	Verbenaceae	Lamiales
Glechoma	ground ivy, creeping charlie	Lamiaceae or Labiatae	Lamiales
Gleditsia	honey locust	Fabaceae or Leguminosae	Fabales
Glehnia	silvertop	Apiaceae	Apiales
Gloriosa	gloriosa lily	Colchicaceae	Liliales
Glottiphyllum	tongueleaf plant	Aizoaceae	Caryophyllales
Glycine	soybean	Fabaceae or Leguminosae	Fabales
Glycyrrhiza	licorice	Fabaceae or Leguminosae	Fabales
Gnaphalium	cudweed, everlasting	Asteraceae or Compositae	Asterales
Gomphrena	globe amaranth	Amaranthaceae	Caryophyllales
Goodyera	rattlesnake plantain	Orchidaceae	Asparagales
Gordonia	loblolly-bay	Theaceae	Ericales
Gossypium	cotton	Malvaceae	Malvales
Graptophyllum	caricature plant	Acanthaceae	Lamiales
Grindelia	gumweed	Asteraceae or Compositae	Asterales
Guaiacum	lignum vitae	Zygophyllaceae	Zygophyllales
Gunnera	gunnera	Gunneraceae	Gunnerales
Guzmania	guzmania	Bromeliaceae	Poales
Gymnocladus	Kentucky coffee tree	Fabaceae or Leguminosae	Fabales
Gypsophila	baby's breath, gypsophila	Caryophyllaceae	Caryophyllales
Hackelia	stickseed	Boraginaceae	not assigned
Hamamelis	witch hazel	Hamamelidaceae	Saxifragales
Hamelia	firebush, firecracker shrub	Rubiaceae	Gentianales
Haplopappus	haplopappus	Asteraceae or Compositae	Asterales
Haworthia	haworthia	Asphodelaceae	Asparagales
Hebe	hebes, New Zealand lilac	Plantaginaceae	Lamiales
Hedera	English ivy	Araliaceae	Apiales
Hedychium	ginger lily	Zingiberaceae	Zingiberales
Hedyotis	bluet, Quaker ladies	Rubiaceae	Gentianales
Helenium	sneezeweed	Asteraceae or Compositae	Asterales
Helianthella	dwarf sunflower	Asteraceae or Compositae	Asterales
Helianthemum	sunrose	Cistaceae	Malvales
Helianthus	sunflower, Jerusalem artichoke	Asteraceae or Compositae	Asterales
Helichrysum	strawflower	Asteraceae or Compositae	Asterales
Heliconia	heliconia, parrot flower	Heliconiaceae	Zingiberales
Heliotropium	heliotrope	Boraginaceae	not assigned
Helleborus	Christmas rose, Lenten rose	Ranunculaceae	Ranunculales
Helonias	swamp pink	Melanthiaceae	Liliales

Hemerocallis	daylily	Hemerocallidaceae	Asparagales
Hemiphylacus	hemiphylacus	Asparagaceae	Asparagales
Heptacodium	seven sons flower	Caprifoliaceae	Dipsacales
Heracleum	cow parsnip	Apiaceae or Umbelliferae	Apiales
Hesperaloe	red yucca	Agavaceae	Asparagales
Hesperis	dame's rocket, rocket	Brassicaceae or Cruciferae	Brassicales
Hesperocallis	desert lily	Agavaceae	Asparagales
Hesperolinon	dwarf flax	Linaceae	Malpighiales
Heteranthera	mud plantain	Pontederiaceae	Commelinales
Heteromeles	toyon	Rosaceae	Rosales
Heterotheca	golden aster	Asteraceae or Compositae	Asterales
Heuchera	coral bells, alum root	Saxifragaceae	Saxifragales
Hevea	rubber tree	Euphorbiaceae	Malpighiales
Hibiscus	hibiscus, rose mallow, rose of Sharon	Malvaceae	Malvales
Hieracium	hawkweed, rattlesnake weed	Asteraceae or Compositae	Asterales
Hillebrandia	Hawaiian begonia	Begoniaceae	Cucurbitales
Hippeastrum	amaryllis	Amaryllidaceae	Asparagales
Hippuris	mare's tail	Plantaginaceae	Lamiales
Hoffmanseggia	hog potato, mesquite weed	Fabaceae or Leguminosae	Fabales
Hohenbergia	bromeliad	Bromeliaceae	Poales
Holodiscus	ocean spray, mountain spray	Rosaceae	Rosales
Hordeum	barley	Poaceae or Gramineae	Poales
Hosta	hosta, plantain lily	Agavaceae	Asparagales
Hottonia	featherfoil, water violet	Primulaceae	Ericales
Houstonia	see *Hedyotis*		
Houttuynia	houttuynia	Saururaceae	Piperales
Hoya	wax plant	Apocynaceae	Gentianales
Hudsonia	beach heather	Cistaceae	Malvales
Humulus	hops	Cannabaceae	Rosales
Hyacinthoides	wood hyacinth	Hyacinthaceae	Asparagales
Hyacinthus	hyacinth	Hyacinthaceae	Asparagales
Hybanthus	green violet	Violaceae	Malpighiales
Hydrangea	hydrangea	Hydrangeaceae	Cornales
Hydrastis	goldenseal	Ranunculaceae	Ranunculales
Hydrocharis	frog's bit	Hydrocharitaceae	Alismatales
Hydrophyllum	waterleaf	Boraginaceae	not assigned
Hylocereus	night-blooming cereus	Cactaceae	Caryophyllales
Hymenocallis	spider lily	Amaryllidaceae	Asparagales
Hymenoxys	rubber sunflower, bitterweed	Asteraceae or Compositae	Asterales
Hyoscyamus	henbane	Solanaceae	Solanales
Hypericum	St. John's wort, Klamath weed	Hypericaceae	Malpighiales
Hypoestes	polka-dot plant	Acanthaceae	Lamiales
Hyssopus	hyssop	Lamiaceae or Labiatae	Lamiales
Iberis	candytuft	Brassicaceae or Cruciferae	Brassicales
Ilex	holly	Aquifoliaceae	Aquifoliales
Illicium	star anise, anise tree	Illiciaceae	Austrobaileyales
Impatiens	impatiens, busy lizzie, touch-me-not, balsam	Balsaminaceae	Ericales
Incarvillea	hardy gloxinia	Bignoniaceae	Lamiales

Indigofera	indigo	Fabaceae or Leguminosae	Fabales
Inula	yellowhead, inula, elecampane	Asteraceae or Compositae	Asterales
Ipheion	spring star flower	Alliaceae	Asparagales
Ipomoea	morning glory, sweet potato	Convolvulaceae	Solanales
Ipomopsis	scarlet gilia, skyrocket	Polemoniaceae	Ericales
Iresine	bloodleaf	Amaranthaceae	Caryophyllales
Iris	iris	Iridaceae	Asparagales
Isatis	dyer's woad	Brassicaceae or Cruciferae	Brassicales
Itea	sweetspire	Iteaceae	Saxifragales
Iva	marshelder	Asteraceae or Compositae	Asterales
Ixia	African corn lily	Iridaceae	Asparagales
Ixora	ixora, flame-of-the-woods	Rubiaceae	Gentianales
Jacaranda	jacaranda tree	Bignoniaceae	Lamiales
Jacquemontia	clustervine	Convolvulaceae	Solanales
Jamesia	waxflower (temperate native)	Hydrangeaceae	Cornales
Jasione	sheep's bit	Campanulaceae	Asterales
Jasminum	jasmine	Oleaceae	Lamiales
Jatropha	coral plant, physic-nut	Euphorbiaceae	Malpighiales
Jeffersonia	twin leaf	Berberidaceae	Ranunculales
Juglans	walnut	Juglandaceae	Fagales
Juncus	rush	Juncaceae	Poales
Justicia	shrimp plant, water willow, Brazilian plume flower	Acanthaceae	Lamiales
Kaempferia	peacock ginger, galanga	Zingiberaceae	Zingiberales
Kalanchoe	kalanchoe, maternity plant, air plant	Crassulaceae	Saxifragales
Kalmia	mountain laurel, alpine laurel	Ericaceae	Ericales
Kerriodoxa	white elephant palm, King Thai palm	Arecaceae	Arecales
Kigelia	sausage tree	Bignoniaceae	Lamiales
Knautia	blue buttons, field scabious	Dipsacaceae	Dipsacales
Kniphofia	red-hot poker	Asphodelaceae	Asparagales
Kobresia	bog sedge	Cyperaceae	Poales
Kochia	kochia weed, molly, summer cypress	Amaranthaceae	Caryophyllales
Koelreuteria	golden rain tree, Chinese flame tree, flamegold	Sapindaceae	Sapindales
Kolkwitzia	beauty bush	Linnaeaceae	Dipsacales
Krameria	rhatany	Krameriaceae	Zygophyllales
Krugiodendron	leadwood	Rhamnaceae	Rosales
Lablab	hyacinth bean	Fabaceae or Leguminosae	Fabales
Lactuca	lettuce, wild lettuce, prickly lettuce	Asteraceae or Compositae	Asterales
Lagenaria	bottle gourd	Cucurbitaceae	Cucurbitales
Lagerstroemia	crape myrtle	Lythraceae	Myrtales
Lamium	dead nettle, henbit	Lamiaceae or Labiatae	Lamiales
Lampranthus	ice plant	Aizoaceae	Caryophyllales
Lantana	lantana	Verbenaceae	Lamiales
Laportea	wood nettle	Urticaceae	Rosales
Lappula	stickseed	Boraginaceae	not assigned

Larrea	creosote bush	Zygophyllaceae	Zygophyllales
Lathyrus	sweet pea	Fabaceae or Leguminosae	Fabales
Laurus	sweet bay tree	Lauraceae	Laurales
Lavadula	lavender	Lamiaceae or Labiatae	Lamiales
Lavatera	tree mallow	Malvaceae	Malvales
Lawsonia	henna	Lythraceae	Myrtales
Ledum	Labrador tea, trapper's tea	Ericaceae	Ericales
Lemna	duckweed	Araceae	Alismatales
Lens	lentil	Fabaceae or Leguminosae	Fabales
Leontopodium	edelweiss	Asteraceae or Compositae	Asterales
Lepidium	pepperweed, peppergrass	Brassicaceae or Cruciferae	Brassicales
Leptospermum	tea tree, manaka	Myrtaceae	Myrtales
Lesquerella	bladderpod	Brassicaceae or Cruciferae	Brassicales
Leucanthemum	Shasta daisy, ox-eye daisy	Asteraceae or Compositae	Asterales
Leucojum	snowflake, snowdrop	Amaryllidaceae	Asparagales
Leucophyllum	silverleaf, Texas ranger	Scrophulariaceae	Lamiales
Levisticum	lovage	Apiaceae or Umbelliferae	Apiales
Lewisia	bitterroot, lewisia	Portulacaceae	Caryophyllales
Liatris	gayfeather, blazing star	Asteraceae or Compositae	Asterales
Ligularia	ligularia, giant groundsel	Asteraceae or Compositae	Asterales
Ligusticum	Porter lovage, oshá	Apiaceae or Umbelliferae	Apiales
Ligustrum	privet	Oleaceae	Lamiales
Lilium	lily, including wood lily, tiger lily, and Easter lily	Liliaceae	Liliales
Limnobium	frog's bit	Hydrocharitaceae	Alismatales
Limonium	statice, sea lavender	Plumbaginaceae	Caryophyllales
Linanthus	linanthus, desert gold, mustang clover, flaxflower	Polemoniaceae	Ericales
Linaria	toadflax	Plantaginaceae	Lamiales
Lindera	spice bush	Lauraceae	Laurales
Linnaea	twinflower	Linnaeaceae	Dipsacales
Linum	flax	Linaceae	Malpighiales
Lippia	lemon verbena	Verbenaceae	Lamiales
Liquidamber	sweet gum tree	Altingiaceae	Saxifragales
Liriodendron	tulip tree, tulip popular	Magnoliaceae	Magnoliales
Liriope	lily turf	Ruscaceae	Asparagales
Listera	twayblade orchid	Orchidaceae	Asparagales
Litchi	lychee	Sapindaceae	Sapindales
Lithocarpus	tanoak	Fagaceae	Fagales
Lithophragma	woodland star	Saxifragaceae	Saxifragales
Lithops	living stones	Aizoaceae	Caryophyllales
Lithospermum	puccoon, gromwell	Boraginaceae	not assigned
Litsea	pond spice	Lauraceae	Laurales
Littonia	climbing lily	Colchicaceae	Liliales
Lobelia	lobelia, cardinal flower	Campanulaceae	Asterales
Lobularia	sweet alyssum	Brassicaceae or Cruciferae	Brassicales
Lomatium	biscuit root	Apiaceae or Umbelliferae	Apiales
Lonicera	honeysuckle	Caprifoliaceae	Dipsacales
Lophophora	peyote	Cactaceae	Caryophyllales

Loropetalum	Chinese witch hazel	Hamamelidaceae	Saxifragales
Ludwigia	seedbox, water primrose	Onagraceae	Myrtales
Luffa	vegetable sponge	Cucurbitaceae	Cucurbitales
Lunaria	honesty, money plant	Brassicaceae or Cruciferae	Brassicales
Lupinus	lupine	Fabaceae or Leguminosae	Fabales
Luzula	wood rush	Juncaceae	Poales
Lychnis	Maltese cross, rose campion	Caryophyllaceae	Caryophyllales
Lycium	desert thorn, wolfberry	Solanaceae	Solanales
Lycopersicon	see *Solanum*		
Lygodesmia	skeleton weed, pink dandelion	Asteraceae or Compositae	Asterales
Lyonia	maleberry, staggerbush	Ericaceae	Ericales
Lysichiton	yellow skunk cabbage (western)	Araceae	Alismatales
Lysimachia	yellow loosestrife, creeping jenny, moneywort, golden globes	Myrsinaceae	Ericales
Lythrum	purple loosestrife	Lythraceae	Myrtales
Macadamia	macadamia nut	Proteaceae	Proteales
Macfadyena	yellow trumpet vine, cat's claw	Bignoniaceae	Lamiales
Machaeranthera	Mojave aster, Tahoka daisy, sticky daisy	Asteraceae or Compositae	Asterales
Macleaya	plume poppy	Papaveraceae	Ranunculales
Maclura	Osage orange, bois d'arc	Moraceae	Rosales
Madia	tarweed	Asteraceae or Compositae	Asterales
Magnolia	magnolia	Magnoliaceae	Magnoliales
Mahonia	Oregon grape holly	Berberidaceae	Ranunculales
Maianthemum	wild lily-of-the-valley	Ruscaceae	Asparagales
Malephora	ice plant	Aizoaceae	Caryophyllales
Malpighia	Barbados cherry	Malpighiaceae	Malpighiales
Malus	apple, crabapple	Rosaceae	Rosales
Malva	mallow, cheeseweed	Malvaceae	Malvales
Malvaviscus	turk's cap, turk's turban	Malvaceae	Malvales
Mammillaria	pincushion cactus, fishhook cactus	Cactaceae	Caryophyllales
Mandevilla	Chilean jasmine, mandevilla	Apocynaceae	Gentianales
Mandragora	mandrake	Solanaceae	Solanales
Mangifera	mango	Anacardiaceae	Sapindales
Manihot	tapioca, cassava	Euphorbiaceae	Malpighiales
Manilkara	sapodilla, chicle	Sapotaceae	Ericales
Marah	wild cucumber, spiny cucumber, manroot	Cucurbitaceae	Cucurbitales
Maranta	prayer plant, arrowroot	Marantaceae	Zingiberales
Marrubium	horehound	Lamiaceae or Labiatae	Lamiales
Marshallia	Barbara's buttons	Asteraceae or Compositae	Asterales
Matelea	milkvine, spinypod	Apocynaceae	Gentianales
Matricaria	pineapple weed, chamomile (tea herb)	Asteraceae or Compositae	Asterales
Matthiola	stocks	Brassicaceae or Cruciferae	Brassicales
Mazus	mazus	Phrymaceae	Lamiales
Meconopsis	Himalayan poppy, Welsh poppy	Papaveraceae	Ranunculales
Medicago	alfalfa	Fabaceae or Leguminosae	Fabales
Melanthium	bunchflower	Melanthiaceae	Liliales

Melia	chinaberry tree	Meliaceae	Sapindales
Melilotus	sweet clover, yellow clover	Fabaceae or Leguminosae	Fabales
Melissa	lemon balm	Lamiaceae or Labiatae	Lamiales
Menispermum	moonseed	Menispermaceae	Ranunculales
Mentha	mint, pennyroyal	Lamiaceae or Labiatae	Lamiales
Mentzelia	blazing star, stick-leaf	Loasaceae	Cornales
Menyanthus	buckbean, bogbean	Menyanthaceae	Asterales
Merremia	Hawaiian wood rose, noyau vine	Convolvulaceae	Solanales
Mertensia	bluebells, mertensia	Boraginaceae	not assigned
Mesembry-anthemum	ice plant	Aizoaceae	Caryophyllales
Metopium	Florida poisonwood	Anacardiaceae	Sapindales
Metrosideros	New Zealand Christmas tree, ratas, lehua	Myrtaceae	Myrtales
Mikania	climbing boneset, climbing hempweed	Asteraceae or Compositae	Asterales
Miltonia	pansy orchid	Orchidaceae	Asparagales
Mimosa	mimosa, sensitive plant	Fabaceae or Leguminosae	Fabales
Mimulus	monkey flower	Phrymaceae	Lamiales
Minuartia	sandwort	Caryophyllaceae	Caryophyllales
Mirabilis	four o'clock	Nyctaginaceae	Caryophyllales
Mitchella	partridge berry, twinberry	Rubiaceae	Gentianales
Mitella	mitrewort	Saxifragaceae	Saxifragales
Moluccella	bells of Ireland	Lamiaceae or Labiatae	Lamiales
Monarda	horsemint, bee balm, wild bergamot	Lamiaceae or Labiatae	Lamiales
Moneses	wood nymph, single delight	Ericaceae	Ericales
Monotropa	Indian pipes	Ericaceae	Ericales
Monstera	split leaf philodendron	Araceae	Alismatales
Montia	montia, miner's lettuce	Portulacaceae	Caryophyllales
Morus	mulberry tree	Moraceae	Rosales
Musa	bananas, plantain (tropical fruit)	Musaceae	Zingiberales
Muscari	grape hyacinth	Hyacinthaceae	Asparagales
Myosotis	forget-me-not	Boraginaceae	not assigned
Myrica	bayberry, wax myrtle, candleberry	Myricaceae	Fagales
Myriophyllum	water milfoil	Haloragaceae	Saxifragales
Myrrhis	sweet cicely	Apiaceae or Umbelliferae	Apiales
Myrtus	myrtle	Myrtaceae	Myrtales
Nama	purple mat, cliff mat	Boraginaceae	not assigned
Nandina	sacred bamboo, heavenly bamboo	Berberidaceae	Myrtales
Narcissus	daffodil, jonquil	Amaryllidaceae	Asparagales
Nelumbo	sacred lotus, American lotus	Nelumbonaceae	Proteales
Nemacladus	thread stem	Campanulaceae	Asterales
Nemesia	nemesia	Scrophulariaceae	Lamiales
Nemophila	baby blue eyes	Boraginaceae	not assigned
Neomarica	fan iris, walking iris	Iridaceae	Asparagales
Neoregelia	heart of flame, blushing bromeliad	Bromeliaceae	Poales
Nepenthes	Asian hanging pitcher plant	Nepenthaceae	Caryophyllales

292

Nepeta	catnip, catmint	Lamiaceae or Labiatae	Lamiales
Nerium	oleander	Apocynaceae	Gentianales
Nicolaia	see *Etlingera*		
Nicotiana	tobacco, tree tobacco	Solanaceae	Solanales
Nierembergia	cup flower	Solanaceae	Solanales
Nigella	love-in-a-mist	Ranunculaceae	Ranunculales
Nolina	beargrass	Ruscaceae	Asparagales
Nothoscordum	false garlic, arrow poison	Alliaceae	Asparagales
Nuphar	pond lily	Nymphaeaceae	Nymphaeales
Nyctagina	devil's bouquet	Nyctaginaceae	Caryophyllales
Nymphaea	water lily	Nymphaeaceae	Nymphaeales
Nymphoides	floating heart	Menyanthaceae	Asterales
Nyssa	tupelo trees, black gum tree	Nyssaceae	Cornales
Obolaria	pennywort	Gentianaceae	Gentianales
Ocimum	basil	Lamiaceae or Labiatae	Lamiales
Oenothera	evening primrose	Onagraceae	Myrtales
Olea	olive	Oleaceae	Lamiales
Olneya	ironwood	Fabaceae or Leguminosae	Fabales
Onopordum	Scotch thistle	Asteraceae or Compositae	Asterales
Opuntia	prickly pear cactus, cholla	Cactaceae	Caryophyllales
Orobanche	broomrape	Orobanchaceae	Lamiales
Oreoxis	alpine parsley	Apiaceae or Umbelliferae	Apiales
Origanum	oregano, marjoram	Lamiaceae or Labiatae	Lamiales
Ornithogalum	star of Bethlehem, pregnant onion	Hyacinthaceae	Asparagales
Orontium	golden club	Araceae	Alismatales
Orthocarpus	see *Castilleja*		
Oryza	rice	Poaceae or Gramineae	Poales
Oscularia	ice plant	Aizoaceae	Caryophyllales
Osmanthus	osmanthus, devilwood, sweet olive	Oleaceae	Lamiales
Osmorhiza	wild sweet cicely	Apiaceae or Umbelliferae	Apiales
Osteospermum	African daisy	Asteraceae or Compositae	Asterales
Ostrya	hop hornbeam	Betulaceae	Fagales
Oxalis	oxalis, wood sorrel, shamrock	Oxalidaceae	Oxalidales
Oxybaphus	umbrella-wort	Nyctaginaceae	Caryophyllales
Oxytropis	locoweed	Fabaceae or Leguminosae	Fabales
Pachypodium	Madagascar palm	Apocynaceae	Gentianales
Pachysandra	Japanese spurge	Buxaceae	Buxales
Pachystachys	golden candle	Acanthaceae	Lamiales
Packera	groundsel, ragwort, butterweed	Asteraceae or Compositae	Asterales
Paeonia	peony	Paeoniaceae	Saxifragales
Panax	ginseng	Araliaceae	Apiales
Pandanus	screw pine, screw palm	Pandanaceae	Pandanales
Pandorea	bower vine, wonga-wonga vine	Bignoniaceae	Lamiales
Panicum	common millet	Poaceae or Gramineae	Poales
Papaver	poppy	Papaveraceae	Ranunculales
Paphiopedilum	slipper orchids	Orchidaceae	Asparagales
Parentucellia	parentucellia, glandweed	Orobanchaceae	Lamiales
Parrotia	parrotia	Hamamelidaceae	Saxifragales

Parthenocissus	Virginia creeper, wood vine, Boston ivy	Vitaceae	Vitales
Pascopyrum	western wheatgrass	Poaceae or Gramineae	Poales
Passiflora	passion flower	Passifloraceae	Malpighiales
Pastinaca	parsnip	Apiaceae or Umbelliferae	Apiales
Patrinia	patrinia, golden lace	Valerianaceae	Dipsacales
Paulownia	empress tree	Paulowniaceae	Lamiales
Paxistima	paxistima	Celastraceae	Celastrales
Pedicularis	lousewort, parrot beak, wood betony, red elephant head, Indian warrior	Orobanchaceae	Lamiales
Pediocactus	mountain ball cactus	Cactaceae	Caryophyllales
Pelargonium	garden geranium, scented geranium	Geraniaceae	Geraniales
Peltandra	arrow arum	Araceae	Alismatales
Pennisetum	pearl millet	Poaceae or Gramineae	Poales
Penstemon	penstemon, beardtongue	Plantaginaceae	Lamiales
Pentas	pentas, Egyptian star cluster	Rubiaceae	Gentianales
Peperomia	peperomia	Piperaceae	Piperales
Perideridia	yampa	Apiaceae or Umbelliferae	Apiales
Perovskia	Russian sage	Lamiaceae or Labiatae	Lamiales
Persea	avocado, red bay tree, silk bay tree	Lauraceae	Laurales
Persicaria	snakeweed, knotweed	Polygonaceae	Caryophyllales
Petalonyx	sandpaper plant	Loasaceae	Cornales
Petasites	western coltsfoot	Asteraceae or Compositae	Asterales
Petrea	queen's wreath	Verbenaceae	Lamiales
Petroselinum	parsley	Apiaceae or Umbelliferae	Apiales
Petunia	petunia	Solanaceae	Solanales
Phacelia	phacelia, desert bluebell, scorpionweed	Boraginaceae	not assigned
Phalaenopsis	moth orchid	Orchidaceae	Asparagales
Phaseolus	bean	Fabaceae or Leguminosae	Fabales
Phellodendron	cork tree	Rutaceae	Sapindales
Philadelphus	mock orange	Hydrangeaceae	Cornales
Philodendron	philodendron	Araceae	Alismatales
Phleum	timothy grass	Poaceae or Gramineae	Poales
Phlomis	Jerusalem sage	Lamiaceae or Labiatae	Lamiales
Phlox	phlox, moss pink	Polemoniaceae	Ericales
Phoenix	date palm	Arecaceae or Palmae	Arecales
Pholisma	sand food	Boraginaceae	not assigned
Phoradendron	Christmas mistletoe, desert mistletoe	Santalaceae	Santalales
Phormium	New Zealand flax	Hemerocallidaceae	Asparagales
Phragmites	reeds, giant reed	Poaceae or Gramineae	Poales
Phryma	lopseed	Phrymaceae	Lamiales
Phyla	wedgeleaf, frogfruit	Verbenaceae	Lamiales
Phyllodoce	pink mountain heather	Ericaceae	Ericales
Phyllostachys	hardy bamboo	Poaceae or Gramineae	Poales
Physalis	ground-cherries, tomatillo, Chinese lantern plant	Solanaceae	Solanales

Physaria	double bladder pod	Brassicaceae or Cruciferae	Brassicales
Physocarpus	ninebark	Rosaceae	Rosales
Physostegia	obedient plant	Lamiaceae or Labiatae	Lamiales
Phytolacca	pokeweed	Phytolaccaceae	Caryophyllales
Pieris	pieris, lily-of-the-valley shrub	Ericaceae	Ericales
Pilea	aluminum plant, artillery plant, clear weed	Urticaceae	Rosales
Pimenta	allspice	Myrtaceae	Myrtales
Pimpinella	anise	Apiaceae or Umbelliferae	Apiales
Pinguicula	butterwort	Lentibulariaceae	Lamiales
Piper	black pepper, kava	Piperaceae	Piperales
Pistacia	pistachio nut, Chinese pistache	Anacardiaceae	Sapindales
Pistia	water lettuce	Araceae	Alismatales
Pisum	peas	Fabaceae or Leguminosae	Fabales
Pittosporum	pittosporum	Pittosporaceae	Apiales
Plagiobothrys	popcorn flower	Boraginaceae	not assigned
Planera	planer tree	Ulmaceae	Rosales
Plantago	plantain (temperate, weedy)	Plantaginaceae	Lamiales
Platanthera	fringed orchids	Orchidaceae	Asparagales
Platanus	plane tree, sycamore, buttonwood tree	Platanaceae	Proteales
Platycodon	balloon flower	Campanulaceae	Asterales
Platystemon	cream cups	Papaveraceae	Ranunculales
Plectranthus	Swedish ivy	Lamiaceae or Labiatae	Lamiales
Pluchea	camphorweed, fleabane	Asteraceae or Compositae	Asterales
Plumbago	Cape plumbago, leadwort	Plumbaginaceae	Caryophyllales
Plumeria	frangipani	Apocynaceae	Gentianales
Poa	bluegrass	Poaceae or Gramineae	Poales
Podophyllum	may-apple	Berberidaceae	Ranunculales
Polanisia	clammyweed	Cleomaceae	Brassicales
Polemonium	Jacob's ladder, sky pilot	Polemoniaceae	Ericales
Polygala	milkwort, snakeroot	Polygalaceae	Fabales
Polygonatum	Solomon's seal	Ruscaceae	Asparagales
Polygonum	knotweed, smartweed, bistort	Polygonaceae	Caryophyllales
Poncirus	hardy orange, trifoliate orange	Rutaceae	Sapindales
Pontederia	pickerel weed	Pontederiaceae	Commelinales
Populus	aspen, cottonwood, poplar	Salicaceae	Malpighiales
Portulaca	rose moss, purslane	Portulacaceae	Caryophyllales
Potamogeton	pondweed	Potamogetonaceae	Alismatales
Potentilla	cinquefoil	Rosaceae	Rosales
Primula	primrose	Primulaceae	Ericales
Proboscidea	unicorn plant, devil's claw	Martyniaceae	Lamiales
Prosopis	mesquite	Fabaceae or Leguminosae	Fabales
Prunella	self-heal, heal-all	Lamiaceae or Labiatae	Lamiales
Prunus	plum, cherry, apricot, peach, nectarine, almond, chokecherry, cherry laurel	Rosaceae	Rosales
Psidium	guava, strawberry guava	Myrtaceae	Myrtales
Psilocarpus	wooly heads	Asteraceae or Compositae	Asterales

Psilostrophe	paperflower	Asteraceae or Compositae	Asterales
Psychotria	ipecac	Rubiaceae	Gentianales
Ptelea	hop tree, wafer ash	Rutaceae	Sapindales
Pterocarya	wingnut	Juglandaceae	Fagales
Pterospora	pinedrops	Ericaceae	Ericales
Pulmonaria	lungwort	Boraginaceae	not assigned
Pulsatilla	pasque flower	Ranunculaceae	Ranunculales
Punica	pomegranate	Lythraceae	Myrtales
Puschkinia	puschkinia	Hyacinthaceae	Asparagales
Pyracantha	firethorn, pyracantha	Rosaceae	Rosales
Pyrola	wintergreen (west coast)	Ericaceae	Ericales
Pyrus	pear	Rosaceae	Rosales
Pyxidanthera	pyxie	Diapensiaceae	Ericales
Quercus	oak	Fagaceae	Fagales
Quincula	Chinese lantern, purple ground cherry	Solanaceae	Solanales
Ranunculus	buttercup	Ranunculaceae	Ranunculales
Raphanus	radish, wild radish	Brassicaceae or Cruciferae	Brassicales
Raphia	raffia palm	Arecaceae or Palmae	Arecales
Ratibida	prairie coneflower, Mexican hat	Asteraceae or Compositae	Asterales
Ravenala	traveler's palm	Strelitziaceae	Zingiberales
Reinwardtia	yellow flax	Linaceae	Malpighiales
Reseda	mignonette, dyer's weed	Resedaceae	Brassicales
Rhamnus	buckthorn, redberry, coffeeberry, cascara sagrada	Rhamnaceae	Rosales
Rheum	rhubarb	Polygonaceae	Caryophyllales
Rhexia	meadow beauty	Melastomataceae	Myrtales
Rhinanthus	yellow rattle, rattlebox	Orobanchaceae	Lamiales
Rhododendron	azalea, rhododendron	Ericaceae	Ericales
Rhus	sumac, skunkbush	Anacardiaceae	Sapindales
Rhynchospora	white-topped sedge	Cyperaceae	Poales
Ribes	gooseberry, current	Grossulariaceae	Saxifragales
Ricinus	castor bean	Euphorbiaceae	Malpighiales
Rivina	rouge plant, pigeonberry	Phytolaccaceae	Caryophyllales
Robinia	locust tree	Fabaceae or Leguminosae	Fabales
Romneya	California tree poppy	Papaveraceae	Ranunculales
Rosa	rose	Rosaceae	Rosales
Rosmarinus	rosemary	Lamiaceae or Labiatae	Lamiales
Rubia	madder	Rubiaceae	Gentianales
Rubus	blackberry, raspberry, bramble, boysenberry, dewberry, cloud-berry, loganberry	Rosaceae	Rosales
Rudbeckia	black-eyed Susan, coneflowers	Asteraceae or Compositae	Asterales
Ruellia	ruellia, wild petunia, willowleaf petunia	Acanthaceae	Lamiales
Rumex	sorrel, dock	Polygonaceae	Caryophyllales
Ruscus	butcher's broom	Ruscaceae	Asparagales
Russelia	coral plant, firecracker plant	Plantaginaceae	Laminales
Ruta	rue, herb of grace	Rutaceae	Sapindales
Sabal	palmetto	Arecaceae	Arecales

296

Sabatia	saltmarsh pink, rose gentian	Gentianaceae	Gentianales
Saccharum	sugar cane	Poaceae or Gramineae	Poales
Sagina	Irish moss, Scotch moss	Caryophyllaceae	Caryophyllales
Sagittaria	arrowhead, wapato	Alismataceae	Alismatales
Saintpaulia	African violet	Gesneriaceae	Lamiales
Sairocarpus	wild snapdragon	Plantaginaceae	Lamiales
Salicornia	glasswort, saltwort	Amaranthaceae	Caryophyllales
Salix	willow	Salicaceae	Malpighiales
Salpiglossis	painted tongue	Solanaceae	Solanales
Salsola	tumbleweed, Russian thistle	Amaranthaceae	Caryophyllales
Salvia	sage	Lamiaceae or Labiatae	Lamiales
Sambucus	elderberry, elder	Adoxaceae	Dipsacales
Sanguinaria	bloodroot	Papaveraceae	Ranunculales
Sanguisorba	burnet	Rosaceae	Rosales
Sanicula	black snakeroot	Apiaceae or Umbelliferae	Apiales
Sansevieria	mother-in-law's tongue	Ruscaceae	Asparagales
Santalum	sandalwood	Santalaceae	Santalales
Santolina	lavender cotton, santolina	Asteraceae or Compositae	Asterales
Sapindus	soapberry	Sapindaceae	Sapindales
Saponaria	soapwort, bouncing bet	Caryophyllaceae	Caryophyllales
Sarcobatus	greasewood	Sarcobataceae	Caryophyllales
Sarracenia	pitcher plant, trumpet pitcher	Sarraceniaceae	Ericales
Saruma	Chinese wild ginger	Aristolochiaceae	Piperales
Sassafras	sassafras	Lauraceae	Laurales
Satureja	savory, yerba buena	Lamiaceae or Labiatae	Lamiales
Saururus	lizard tail	Saururaceae	Piperales
Saxifraga	saxifrage, strawberry begonia	Saxifragaceae	Saxifragales
Scabiosa	pincushion flower	Dipsacaceae	Dipsacales
Schefflera	umbrella plant	Araliaceae	Apiales
Schinus	California peppertree	Anacardiaceae	Sapindales
Schisandra	wild sarsaparilla, star vine	Schisandraceae	Austrobaileyales
Schizanthus	butterfly flower, poor man's orchid	Solanaceae	Solanales
Schlumbergera	Christmas cactus, crab cactus	Cactaceae	Caryophyllales
Schoenoplectus	tule, bulrush	Cyperaceae	Poales
Scilla	squill	Hyacinthaceae	Asparagales
Scindapus	satin pothos	Araceae	Alismatales
Scirpus	bulrush, wool grass	Cyperaceae	Poales
Scropularia	figwort, carpenter's square	Scrophulariaceae	Lamiales
Scutellaria	skullcap	Lamiaceae or Labiatae	Lamiales
Secale	rye	Poaceae or Gramineae	Poales
Sechium	chayote	Cucurbitaceae	Cucurbitales
Sedum	stonecrop, sedum	Crassulaceae	Saxifragales
Selinocarpus	monopod	Nyctaginaceae	Caryophyllales
Sempervivum	hen and chickens, houseleek	Crassulaceae	Saxifragales
Senecio	senecio, groundsel, ragwort, butterweed, dusty miller, cineraria	Asteraceae or Compositae	Asterales
Senna	senna	Fabaceae or Leguminosae	Fabales
Sesamum	sesame	Pedaliaceae	Lamiales

Shepherdia	buffaloberry	Elaeagnaceae	Rosales
Shortia	Oconee bells, fringe bells	Diapensiaceae	Ericales
Sicyos	bur cucumber, star cucumber	Cucurbitaceae	Cucurbitales
Sidalcea	checker mallow, miniature hollyhock	Malvaceae	Malvales
Silene	moss campion, cushion pink, fire pink, silene	Caryophyllaceae	Caryophyllales
Silphium	compass plant, rosin weed	Asteraceae or Compositae	Asterales
Silybum	milk thistle	Asteraceae or Compositae	Asterales
Simarouba	paradise tree	Simaroubaceae	Sapindales
Simmondsia	jojoba	Simmondsiaceae	Caryophyllales
Sinningia	gloxinia, cardinal flower	Gesneriaceae	Lamiales
Sisymbrium	mustard weed, tumble mustard	Brassicaceae or Cruciferae	Brassicales
Sisyrinchium	blue-eyed grass	Iridaceae	Asparagales
Sium	water parsnip	Apiaceae or Umbelliferae	Apiales
Skimmia	skimmia	Rutaceae	Sapindales
Smilacina	see *Maianthemum*		
Smilax	catbrier, greenbrier, bull brier, carrion flower, sarsaparilla	Smilacaceae	Liliales
Solanum	tomato, potato, eggplant, night-shade, horsenettle, potato vine, pepino, Jerusalem cherry, buffa-lobur, European bittersweet	Solanaceae	Solanales
Soleirolia	baby's-tears	Urticaceae	Rosales
Solenostemon	coleus, painted nettle	Lamiaceae or Labiatae	Lamiales
Solidago	goldenrod	Asteraceae or Compositae	Asterales
Sonchus	sow thistle	Asteraceae or Compositae	Asterales
Sophora	necklace pod, mescal bean tree	Fabaceae or Leguminosae	Fabales
Sorbaria	false spirea	Rosaceae	Rosales
Sorbus	mountain ash	Rosaceae	Rosales
Sorghum	sorghum, Johnson grass	Poaceae or Gramineae	Poales
Sparaxis	harlequin flower	Iridaceae	Asparagales
Sparganium	bur-reed	Sparganiaceae	Poales
Spartium	Spanish broom	Fabaceae or Leguminosae	Fabales
Spathiphyllum	peace lily	Araceae	Alismatales
Sphaeralcea	globe mallow	Malvaceae	Malvales
Spigelia	Indian pink, pink root	Loganiaceae	Gentianales
Spinacia	spinach	Amaranthaceae	Caryophyllales
Spiranthes	lady's tresses	Orchidaceae	Asparagales
Spiraea	spiraea, bridal wreath, meadow-sweet	Rosaceae	Rosales
Sprekelia	Jacobean lily, Aztec lily	Amaryllidaceae	Asparagales
Stachys	lamb's ears, betony, hidalgo	Lamiaceae or Labiatae	Lamiales
Stapelia	carrion flower, starfish flower	Apocynaceae	Gentianales
Stellaria	chickweed	Caryophyllaceae	Caryophyllales
Stephanotis	Madagascar jasmine	Apocynaceae	Gentianales
Stevia	stevia, sweet herb, candyleaf	Asteraceae or Compositae	Asterales
Stewartia	stewartia	Theaceae	Ericales
Strelitzia	bird-of-paradise	Strelitziaceae	Zingiberales
Streptopus	twisted stalk	Liliaceae	Liliales

Streptocarpus	cape primrose	Gesneriaceae	Lamiales
Striga	witchweed	Orobanchaceae	Lamiales
Stromanthe	stromanthe, red rain	Marantaceae	Zingiberales
Stylocline	neststraw	Asteraceae or Compositae	Asterales
Stylophorum	wood poppy	Papaveraceae	Ranunculales
Styrax	snowbell	Styracaceae	Ericales
Swertia	star gentian	Gentianaceae	Gentianales
Swietenia	mahogany	Meliaceae	Sapindales
Symphoricarpos	snowberry, coralberry	Caprifoliaceae	Dipsacales
Symphytum	comfrey	Boraginaceae	not assigned
Symplocarpus	skunk cabbage (eastern)	Araceae	Alismatales
Symplocos	sapphireberry, sweet leaf	Symplocaceae	Ericales
Syngonium	arrowhead vine	Araceae	Alismatales
Syringa	lilac	Oleaceae	Lamiales
Syzygium	clove	Myrtaceae	Myrtales
Tabebuia	trumpet tree	Bignoniaceae	Lamiales
Tacca	bat flower	Dioscoreaceae	Dioscoreales
Tagetes	marigold	Asteraceae or Compositae	Asterales
Tamarindus	tamarind	Fabaceae or Leguminosae	Fabales
Tamarix	salt cedar, tamarix	Tamaricaceae	Caryophyllales
Tanacetum	tansy, feverfew, costmary	Asteraceae or Compositae	Asterales
Taraxacum	dandelion	Asteraceae or Compositae	Asterales
Tecoma	orange bells, cape honeysuckle	Bignoniaceae	Lamiales
Tectona	teak	Lamiaceae or Labiatae	Lamiales
Tellima	fringecup	Saxifragaceae	Saxifragales
Tetradymia	horsebrush, cottonthorn	Asteraceae or Compositae	Asterales
Tetragonia	New Zealand spinach	Aizoaceae	Caryophyllales
Tetraneuris	alpine sunflower, perky Sue	Asteraceae or Compositae	Asterales
Tetrapanax	Chinese rice paper plant	Araliaceae	Apiales
Teucrium	germander, cat thyme	Lamiaceae or Labiatae	Lamiales
Thalia	fire flag, water canna	Marantaceae	Zingiberales
Thalictrum	meadow rue	Ranunculaceae	Ranunculales
Theobroma	cacao, chocolate	Malvaceae	Malvales
Thermopsis	golden peas, golden banner	Fabaceae or Leguminosae	Fabales
Thlapsi	field pennycress	Brassicaceae or Cruciferae	Brassicales
Thunbergia	black-eyed susan vine	Acanthaceae	Lamiales
Thymus	thyme	Lamiaceae or Labiatae	Lamiales
Tiarella	foamflower	Saxifragaceae	Saxifragales
Tigridia	Mexican shell flower	Iridaceae	Asparagales
Tilia	linden tree, basswood tree, lime	Malvaceae	Malvales
Tillandsia	Spanish moss, pink quill	Bromeliaceae	Poales
Titanopsis	concrete leaf, little tortoise foot	Aizoaceae	Caryophyllales
Tithonia	Mexican sunflower	Asteraceae or Compositae	Asterales
Tolmiea	piggy-back plant	Saxifragaceae	Saxifragales
Townsendia	Easter daisy, townsendia	Asteraceae or Compositae	Asterales
Toxicodendron	see *Rhus*		
Trachelium	trachelium, throatwort	Campanulaceae	Asterales
Tradescantia	spiderworts, wandering Jew, inch plant, chain plant	Commelinaceae	Commelinales

Tragopogon	salsify, oyster plant, goatsbeard	Asteraceae or Compositae	Asterales
Tribulus	puncture vine, goathead	Zygophyllaceae	Zygophyllales
Trichostema	blue curls, vinegar weed	Lamiaceae or Labiatae	Lamiales
Tricyrtis	toad lily	Liliaceae	Liliales
Trientalis	starflower	Myrsinaceae	Ericales
Trifolium	clover	Fabaceae or Leguminosae	Fabales
Trillium	trilliums, wake robin	Melanthiaceae	Liliales
Triodanis	Venus' looking glass	Campanulaceae	Asterales
Triosteum	wild coffee, feverwort	Caprifoliaceae	Dipsacales
Tripterocalyx	sand puffs	Nyctaginaceae	Caryophyllales
Triticum	wheat, spelt	Poaceae or Gramineae	Poales
Tritonia	tritonia	Iridaceae	Asparagales
Trollius	globeflower	Ranunculaceae	Ranunculales
Tropaeolum	nasturtium, canary creeper	Tropaeolaceae	Brassicales
Tulbaghia	society garlic	Alliaceae	Asparagales
Tulipa	tulip	Liliaceae	Liliales
Tussilago	coltsfoot	Asteraceae or Compositae	Asterales
Typha	cattail	Typhaceae	Poales
Ulmus	elm	Ulmaceae	Rosales
Umbellularia	California bay tree, California laurel, Oregon myrtle	Lauraceae	Laurales
Uniola	sea oats	Poaceae or Gramineae	Poales
Urtica	stinging nettle	Urticaceae	Rosales
Utricularia	bladderwort	Lentibulariaceae	Lamiales
Uvularia	bellwort	Colchicaceae	Liliales
Vaccaria	cowcockle	Caryophyllaceae	Caryophyllales
Vaccinium	blueberry, huckleberry, cranberry	Ericaceae	Ericales
Valeriana	valerian, garden heliotrope	Valerianaceae	Dipsacales
Vallisneria	tape grass	Hydrocharitaceae	Alismatales
Vancouveria	inside-out flower, vancouveria	Berberidaceae	Ranunculales
Vanda	lei orchid	Orchidaceae	Asparagales
Vanilla	vanilla orchid	Orchidaceae	Asparagales
Veratrum	corn husk lily, white hellebore	Melanthiaceae	Liliales
Verbascum	mullein	Scrophulariaceae	Lamiales
Verbena	verbena, vervain	Verbenaceae	Lamiales
Veronica	speedwell, veronicas, brooklime	Plantaginaceae	Lamiales
Viburnum	viburnum, snowball bush, snowball tree, arrowwood, nannyberry, European cranberry, black haw	Adoxaceae	Dipsacales
Vicia	vetch	Fabaceae or Leguminosae	Fabales
Victoria	Victorian water lily	Nymphaeaceae	Nymphaeales
Vinca	periwinkle	Apocynaceae	Gentianales
Viola	violet, viola, pansy, Johnny-jump-up	Violaceae	Malpighiales
Vitex	chaste tree	Lamiaceae or Labiatae	Lamiales
Vitis	grape vine	Vitaceae	Vitales
Vriesea	flaming sword bromeliad, lobster claw bromeliad	Bromeliaceae	Poales
Wahlenbergia	royal bluebell	Campanulaceae	Asterales

Wasabia	wasabi	Brassicaceae or Cruciferae	Brassicales
Washingtonia	California fan palm, Mexican fan palm	Arecaceae or Palmae	Arecales
Weigela	weigela	Diervillaceae	Dipsacales
Whipplea	modesty, whipplea	Hydrangeaceae	Cornales
Wisteria	wisteria	Fabaceae or Leguminosae	Fabales
Wolffia	watermeal	Araceae	Alismatales
Wyethia	mule's ears	Asteraceae or Compositae	Asterales
Xanthium	cocklebur	Asteraceae or Compositae	Asterales
Xanthosoma	tannia, yautia	Araceae	Alismatales
Xeranthemum	immortelle	Asteraceae or Compositae	Asterales
Xyris	yellow-eyed grass	Xyridaceae	Poales
Yucca	yucca, Joshua tree	Agavaceae	Asparagales
Zantedeschia	calla lily	Araceae	Alismatales
Zanthoxylum	prickly-ash, Hercules-club	Rutaceae	Sapindales
Zauschneria	see *Epilobium*		
Zea	corn, popcorn, sweet corn	Poaceae or Gramineae	Poales
Zebrina	see *Tradescantia*		
Zelkova	sawleaf zelkova, Japanese zelkova	Ulmaceae	Rosales
Zephyranthes	zephyr lily, atamasco lily	Amaryllidaceae	Asparagales
Zigadenus	wand lily or death camas	Melanthiaceae	Liliales
Zingiber	ginger	Zingiberaceae	Zingiberales
Zinnia	zinnia	Asteraceae or Compositae	Asterales
Zizania	wild rice	Poaceae or Gramineae	Poales
Zizia	golden alexanders	Apiaceae or Umbelliferae	Apiales
Ziziphus	jujube, gray thorn	Rhamnaceae	Rosales
Zostera	eelgrass	Zosteraceae	Alismatales

Glossary

achene: a closed, dry fruit with one seed; the seed attaches to the relatively thin hull at one point. Example: sunflower seed.

anther: the part of a stamen where the pollen develops and is released. It is attached to the top of the filament.

aromatic: fragrant, having a marked, usually pleasant smell.

asymmetric: shaped such that it cannot be divided into two equal halves.

berry: a simple, fleshy fruit that usually has several seeds and develops from a single pistil. Examples: watermelon, blueberry.

bilateral: having two equal sides; can be divided into two equal parts with a cut in only one place. Example: snapdragon flower.

bisexual: having both stamens and carpels (pistils) – the most common structure for flowers.

bract: a modified leaf that is usually smaller than the normal leaves and is often found on the floral shoot.

bulb: an underground bud with thickened, fleshy modified leaves.

calyx: all of the sepals together; the outermost whorl of flower parts and of the perianth.

capsule: a dry fruit that forms from a pistil with two or more carpels and that splits open when it is mature.

carpel: a simple pistil; the unit of compound pistils; the ovule-bearing unit of the flower, consisting of the ovary, style, and stigma.

carpellate: term used to tell how many units make up a pistil. Example: the three-carpellate pistil of a lily is made of three carpels fused together.

carpellate flower: a flower that has one or more carpels, but no stamens. Also called a pistillate flower; casually called a female flower.

claw: the markedly narrowed base portion of a petal. A clawed petal is one that has a claw. Example: carnation petal.

cladophyll: a stem that has the form and function of a leaf.

compound: composed of two or more similar units.

compound leaf: a leaf that has two or more blades; it is rarely found outside the eudicots.

contractile roots: modified roots that pull the shoot or bulb deeper into the ground.

cordate: heart-shaped. Usually describes the shape of leaves.

corm: a modified stem that is short, fleshy, and underground. Example: gladiolus corm.

corolla: all of the petals of a flower taken together; the inner whorl of the perianth.

corona: an outgrowth of the corolla or stamens, usually showy. Example: daffodil trumpet.

cotyledon: a seed leaf, the first leaf or leaves of the embryo plant.

cyathium: a specialized inflorescence composed of unisexual flowers enclosed in a cup-like structure with nectar glands; characteristic of the Euphorbiaceae. (plural—cyathia)

dioecious: literally, "in two houses"; a species that has unisexual flowers with the sexes on separate plants. Example: holly.

disk flower: a radially symmetrical flower that makes up part or all of the head inflorescence of the family Asteraceae.

distinct: separate, not joined to a like part, not fused.

drupe: a simple, fleshy fruit that has a stony pit. Examples: peach, cherry.

endocarp: the inner layer of the pericarp, usually in a fleshy fruit. The pit of a peach and the gel-like layer around tomato seeds are endocarps.

epicalyx: bracts that lie beneath the true calyx.

ethereal oils: aromatic, oily compounds made by plants and often found in glands or dots on the leaves.

exocarp: the outer layer of a pericarp, usually in a fleshy fruit, like the skin of a peach.

filament: the stalk of the stamen, which supports the anther.

floral shoot: a branch on which one or more flowers grow.

floral cup: a cup-like structure formed from the fusion of the perianth and stamen parts. Example: the base of a rose. Also called a hypanthium.

floral tube: a tube formed from the fusion of perianth and stamen parts. Example: section between the ovary and calyx of evening primrose. Also called a hypanthium.

follicle: a dry fruit that develops from a single carpel and splits open along one side when it is mature.

fruit: the mature, ripened ovary of a pistil. Fruits can be dry seed pods or fleshy and juicy.

fused: grown as if the parts were melted together.

haustoria: roots that are modified to grow into another plant and remove water and nutrients. (singular—haustorium)

head: an inflorescence composed of small flowers attached directly to a short, broad, often disk-like receptacle.

inferior ovary: an ovary that is positioned beneath the perianth and stamens. Examples: squash, daffodil.

inflorescence: two or more flowers on one flower-bearing stem or floral shoot.

internode: the part of a stem that is between two nodes.

involucre: many, closely set bracts that grow beneath a flower or inflorescence. Found in Asteraceae and other families, and often mistaken for a calyx.

legume: a dry fruit that develops from a superior ovary and splits into two halves with seeds on both halves in alternating positions. Example: pea pod.

lobed leaf: a leaf with large, rounded projections from the margin. Example: classic oak leaf.

locule: a chamber or compartment within the ovary, in which the ovules attach. Anthers also have locules, in which the pollen forms.

margin: the outer edge of a plant structure, such as the edge of a leaf blade.

mesocarp: the middle layer of a pericarp, usually developed in a fleshy fruit. Example: the flesh of a peach.

monoecious: literally, "in one house"; a species that has unisexual flowers with both sexes on one plant. Example: squash.

multiple fruit: a fruit formed by the fusion of the separate ovaries from many flowers. Examples: pineapple, mulberry.

mycorrhizae: mutually beneficial partnership between fungi and the roots or other underground parts of a plant. Probably the most widespread symbiosis on the Earth.

nectary: the location in which nectar is produced and presented to pollinators.

node: the place on a stem where a leaf or lateral stem attaches.

numerous: more than about 15 of a part; for stamens, more than twice the number of petals.

nut: a dry fruit that forms from a single carpel and does not split open at maturity. It usually has a thick, woody shell. Example: acorn.

odd-pinnately compound: a compound leaf that has an odd number of leaflets and ends with a single leaflet.

ovary: the part of the pistil or carpel that holds the ovules and develops into the fruit.

ovule: an immature seed. It can become the seed after it is fertilized by cells from pollen.

palmate: radiating out from a single point or area. It can describe leaflets in a compound leaf or a venation pattern.

pappus: the modified sepals of the family Asteraceae, which are bristles, scales, hairs, or fibers.

perennial: a plant that lives for more than two years.

perianth: the two whorls of flower parts beneath the stamens and pistil; the term for the calyx and corolla together.

pericarp: the wall of a fruit. It forms from the wall of the ovary and consists of everything from the seed(s) to the outside of the fruit. See also endocarp, mesocarp, and exocarp.

persistent: remaining attached after its function has been served; refers to sepals or petals that stay on the flower after it has bloomed

petiole: the stalk of a leaf.

phyllary: one of the bracts that make the involucre of Asteraceae flowers. Example: the leaf-like part we pull off from an artichoke when we eat it.

pinnate: with one central stem or vessel and many side branches, like a feather or the teeth of a comb. It can describe leaflets of a compound leaf or a venation pattern. "Bipinnate" describes a compound leaf whose side branches have their own side branches.

pistil: the ovule-bearing part of the flower that consists of one carpel, or two or more fused carpels.

pistillate: a flower with only a pistil or whose only functional reproductive part is a pistil; casually called a female flower.

pome: a fruit of the rose family whose fleshy part is derived from the wall of the receptacle cup. Examples: apple, pear, quince.

raceme: an inflorescence that has a single, elongated stem and in which the flowers each have a short stem and mature from the bottom up.

radial symmetry: structured such that it can be cut in two equal halves in several planes. Example: a petunia flower.

ray flower: a flower of the Asteraceae that has an elongated, strap-like corolla. It can occur around the outside of disk flowers or be the only component of a head. It is often mistaken for petals.

receptacle: the end of the flower-bearing stem where the flower parts attach. It is often broader than the stem.

rhizome: a underground stem that grows horizontally and often functions as storage for food or water.

root nodules: swelling or small knob on a root in which nitrogen-fixing bacteria live. These structures are common in the family Fabaceae, but also are found in other families.

rosette: a dense, rounded cluster of leaves that grows at ground level.

samara: a closed, dry fruit with a wing-like structure that helps it blow away on the wind.

scape: an erect stem that has no leaves and ends in a flower or inflorescence.

schizocarp: a dry fruit from a pistil that has two or more carpels; it breaks into two or more one-seeded, closed segments at maturity. Examples: hollyhock, maple.

serrate leaf: a leaf with forward-pointing teeth on its margin. Biserrate leaves have coarse teeth that are covered with fine teeth.

sessile: literally, "sitting upon"; a stem that is so short it cannot be seen; a leaf that has no petiole.

sheathing leaf: a leaf whose petiole base wraps around the stem.

simple leaf: a leaf with only one blade; not divided into leaflets.

solitary: one at a time; describes single flowers on a floral shoot versus inflorescences

spike: an inflorescence with a single, elongated stem to which the flowers are directly attached. Flowers mature from the bottom upward.

spikelet: a small spike composed of one or more flowers with two bracts beneath them; the basic inflorescence unit of grasses and sedges.

spine: a modified leaf or stipule that is sharp and pointy.

stamen: the pollen-bearing part of the flower, consisting of a filament and an anther.

staminate: a flower with only stamens or whose only functional reproductive parts are stamens.

stigma: the surface at the end of the pistil to which the pollen sticks.

stipule: usually one of a pair of leaf blade–like structures that grow either at the base of the petiole or on the internode; may also be single.

style: the long, narrow part that connects the carpel's or pistil's ovary to its stigma.

succulent: thick and juicy.

superior ovary: an ovary that is positioned on top of the perianth and stamens.

tendril: a modified leaf or stem that is long and thin and that supports a climbing plant by wrapping around structures.

tepal: a perianth segment that is neither petal or sepal; a perianth segment that looks much like all the other segments and serves the same function.

tuber: a short, thick, underground stem that is modified to store food. Example: potato.

umbel: an inflorescence with many branches that are about the same length and that each end in a flower. In compound umbels, the branches end in another smaller umbel.

unisexual: a flower that has either stamens or pistils, but not both.

vascular bundles: groups of xylem and phloem tubes.

venation: the pattern of veins in a leaf. Venation may be pinnate, palmate, or parallel.

whorl: three or more parts of the same kind that are attached at one level around the stem.

Selected References

The following references were consulted in the preparation of this book.

Principal resources:

Stevens, P. F. (2001 onwards). Angiosperm Phylogeny Website. Version 6, May 2005 [and more or less continuously updated since]. Available at: http://www.mobot.org/MOBOT/research/APweb/. Accessed January 6, 2006.

Angiosperm Phylogeny Group. 2003. An update of the Angiosperm Phylogeny Group classification for the orders and families of flowering plants: APG II. Bot. J. Linn. Soc. 141: 399–436.

Funk, V. A., R. J. Bayer, S. Keeley, R. Chan, L. Watson, B. Gemeinholzer, E. Schilling, J. L. Panero, B. G. Baldwin, N. Garcia-Jacas, A. Susanna & R. K. Jansen. 2005. Everywhere but Antarctica: Using a supertree to understand the diversity and distribution of the Compositae. Biol. Skr. 55: 343–374.

Judd, W. S., C. Campbell, E. A. Kellogg, P. F. Stevens & M. J. Donoghue. 2002. Plant Systematics: A Phylogenetic Approach. Sinauer Associates, Sunderland, Massachusetts. [ISBN 0-87893-403-0]

Judd, W. S. & R. G. Olmstead. 2004. A survey of tricolpate (eudicot) phylogenetic relationships. Amer. J. Bot. 91: 1627–1644.

Mabberley, D. J. 1997. The Plant-Book: A Portable Dictionary of the Vascular Plants, ed. 2. Cambridge Univ. Press, Cambridge. [ISBN 0-521-41421-0]

Soltis, P. S. & D. E. Soltis. 2004. The origin and diversification of the angiosperms. Amer. J. Bot. 91: 1614–1626.

Soltis, D. E., P. S. Soltis, P. K. Endress & M. W. Chase. 2005. Phylogeny and Evolution of the Angiosperms. Sinauer Associates, Sunderland, Massachusetts. [ISBN 0-87893-817-6]

Zomlefer, W. 1994. Guide to the Flowering Plant Families. Univ. of North Carolina Press, Chapel Hill. [ISBN 0-8078-4470-5]

Additional resources:

Anderson, E. F. 2001. The Cactus Family. Timber Press, Portland, Oregon. [ISBN 0-88192-498-9]

Angiosperm Phylogeny Group. 1998. An ordinal classification for the families of flowering plants. Ann. Missouri Bot. Gard. 85: 531–553.

Bender, S. (editor) 2004. Southern Living Garden Book. Oxmoor House, Des Moines, Iowa. [ISBN 0-376-03910-8]

Bremer, K. 1994. Asteraceae: Cladistics & Classification. Timber Press, Portland, Oregon. [ISBN 0-88192-275-7]

Bryant, G., T. Rodd & G. von Berg (editors). 1999. Botanica's Annuals and Perennials. Laurel Glenn Publishing, San Diego, California. [ISBN 1-57145-648-1]

Carter, J. L. 1988. Trees and Shrubs of Colorado. Johnson Books, Distributor, Boulder, Colorado. [ISBN 0-9619945-9]

Carter, J. L. 1997. Trees and Shrubs of New Mexico. Johnson Books, Distributor, Boulder, Colorado. [ISBN 0-9658404-0-9]

D'Amato, P. 1998. The Savage Garden: Cultivating Carnivorous Plants. Ten Speed Press, Berkeley, California. [ISBN 0-89815-915-6]

Ellis, M. A. & L. Horst. White pine blister rust on currants and gooseberries. Available at: http://ohioline.osu.edu/hyg-fact/3000/3205.html. Accessed December 29, 2005.

Flora of North America website. Available at: http://www.efloras.org/flora_page.aspx?flora_id=1. Accessed December 29, 2005.

Flora's Plant Names. Timber Press, Portland, Oregon. [ISBN 0-88192-605-1]

Guennel G. K. 2004. Guide to Colorado Wildflowers. Vol. 1. Plains and Foothills, ed. 2. Westcliffe Publishers, Englewood, Colorado. [ISBN 1-56579-512-1]

Guennel G. K. 1995. Guide to Colorado Wildflowers. Vol. 2. Mountains. Westcliffe Publishers, Englewood, Colorado. [ISBN 1-56579-119-3]

Halpin, A. (editor) 2001. Sunset Northeastern Garden Book. Sunset Books, Menlo Park, California. [ISBN 0-376-03524-2]

Heywood, V. H. (editor) 1993. Flowering Plants of the World. Oxford Univ. Press, Oxford. [ISBN 0-19-521037-9]

Hogan, E. L. (editor) 1988. Sunset Western Gardening Book. Sunset Books, Menlo Park, California. [ISBN 0-376-03891-8]

International Plant Names Index (2004). Available at: http://www.ipni.org/index.html. Accessed January 6, 2006.

Integrated Taxonomic Information System (ITIS). Available at: http://www.itis.usda.gov/. Accessed December 29, 2005.

Ivey, R. D. 2003. Flowering Plants of New Mexico, ed. 4. RD & V Ivey Publishers, Albuquerque, New Mexico. [ISBN 0-9612170-3-0]

Kane, D. 2005. Grass-eating dinosaurs discovered. Available at: http://www.aaas.org/news/releases/2005/1117grassdinos.shtml. Accessed December 29, 2005.

Kershaw, L., A. MacKinnon & J. Pojar. 1998. Plants of the Rocky Mountains. Lone Pine Publishing, Edmonton, Alberta, Canada. [ISBN 1-55105-088-9]

Kirkpatrick, Z. M. 1992. Wildflowers of the Western Plains. Univ. of Texas Press, Austin.   [ISBN 0-292-79062-7]

Loughmiller, C. & L. Loughmiller. 1984. Texas Wildflowers: A Field Guide. Univ. of Texas Press, Austin. [ISBN 0-292-78060-5]

Macoboy, S. 2000. What Flower Is That? Chartwell Books, Secaucus, New Jersey. [ISBN 0-7858-1187-7]

McGee, H. 2004. On Food and Cooking: The Science and Lore of the Kitchen. Revised edition. Scribner, New York. [ISBN 0-684-80001-2]

McHoy, P. 1995. The Complete Houseplant Book. Smithmark Books, New York. [ISBN 0-8317-1175-2]

Newcome, L. 1977. Newcomb's Wildflower Guide. Little, Brown and Company, New York. [ISBN 0-316-60442-9]

Phillips, R. & M. Rix. 2002. The Botanical Garden. Vol I and II. Firefly Books, Tonawanda, New York. [ISBN 1-55297-591-6; ISBN 1-55297-592-4]

Phillips, S. J. & P. W. Comus (editors). 2000. A Natural History of the Sonoran Desert. Arizona-Sonora Desert Museum Press, Tucson. [ISBN 0-520-21980-5]

Pojar, J. & A. MacKinnon (editors). 1994. Plants of the Pacific Northwest Coast. Lone Pine Publishing, Edmonton, Alberta, Canada. [ISBN 1-55105-040-4]

Quinlan, K. 2000. Identifying Plants by Family and Genus: Wildflowers and Shrubs of the Western United States. Xlibris, Philadelphia. [ISBN 0-7388-2746-0]

Rieger, M. Cashew – Anacardium occidentale. Available at: http://www.uga.edu/fruit/cashew.htm. Accessed December 29, 2005.

Spellenberg, R. 1979. Audubon Society Field Guide to North American Wildflowers, Western Region. Alfred A. Knopf, New York. [ISBN 0-394-50431-3]

Taylor, R. J. 1990. Northwest Weeds: The Ugly and Beautiful Villains of Fields, Gardens, and Roadsides. Mountain Press Publishing Company, Missoula, Montana. [ISBN 0-87842-249-8]

Thieret, J. W. 2001. Audubon Society Field Guide to North American Wildflowers, Eastern Region. Revised edition. Alfred A. Knopf, New York. [ISBN 0-375-40232-2]

Vaughan, J. G. & C. A. Geissler. 1997. The New Oxford Book of Food Plants. Oxford Univ. Press, Oxford. [ISBN 0-19-850567-1]

United States Department of Agriculture–Agricultural Research Service website. 2000. Watch out water hyacinth. Available at: http://www.ars.usda.gov/is/AR/archive/mar00/water0300.htm. Accessed December 29, 2006.

United States Department of Agriculture–Natural Resources Conservation Service. 2005. The PLANTS Database, Version 3.5 (http://plants.usda.gov). Data compiled from various sources by M. W. Skinner. National Plant Data Center, Baton Rouge, Louisiana. Accessed January 6, 2006.

United States Geological Survey–Northern Rocky Mountain Science Center. Blister rust wars in western national parks. Available at: http://www.nrmsc.usgs.gov/research/blister.htm. Accessed December 29, 2005.

W3TROPICOS website. Available at: http://mobot.mobot.org/W3T/Search/vast.html. Accessed January 6, 2006.

Wasson, E. (editor). 2003. The Complete Encyclopedia of Trees and Shrubs. Thunder Bay Press, San Diego, California. [ISBN 1-59223-055-5]

Willis, K. J. & J. C. McElwain. 2002. The Evolution of Plants. Oxford Univ. Press, Oxford. [ISBN 0-19-850065-3]

Weber, W. A. & R. C. Wittmann. 2001. Colorado Flora: Eastern Slope, ed. 3. Univ. Press of Colorado, Boulder. [ISBN 0-87081-552-0]

Whitson, T. D. (editor). 2000. Weeds of the West, ed. 9. Western Society of Weed Science, Newark, California. [ISBN 0-941570-13-4]

Index of Flowering Plant Families

Entries in bold are illustrated